THE MARINE ALGAL VEGETATION
OF ST. MARTIN, ST. EUSTATIUS AND SABA

THE MARINE ALGAL VEGETATION OF ST. MARTIN, ST. EUSTATIUS AND SABA (NETHERLANDS ANTILLES)

PROEFSCHRIFT

TER VERKRIJGING VAN DE GRAAD VAN DOCTOR
IN DE WISKUNDE EN NATUURWETENSCHAPPEN
AAN DE RIJKSUNIVERSITEIT TE UTRECHT, OP
GEZAG VAN DE RECTOR MAGNIFICUS, PROF. DR.
J. LANJOUW, VOLGENS BESLUIT VAN DE SENAAT
IN HET OPENBAAR TE VERDEDIGEN OP MAANDAG
4 NOVEMBER 1968 DES NAMIDDAGS TE 4 UUR
(PRECIES)

DOOR

M. VROMAN

GEBOREN OP 28 FEBRUARI 1927 TE WOERDEN

1968
Springer-Science+Business Media, B.V.

Promotor: Prof. dr. J. Lanjouw

Dit proefschrift wordt gepubliceerd met financiële steun van de
Landsregering der Nederlandse Antillen en verschijnt tevens als
Uitgaven Natuurwetenschappelijke Studiekring voor Suriname
en de Nederlandse Antillen, No. 52.

ISBN 978-94-017-5660-0   ISBN 978-94-017-5942-7 (eBook)
DOI 10.1007/978-94-017-5942-7

Softcover reprint of the hardcover 1st edition 1968

STELLINGEN

I

De zonering in de litorale en sublitorale regio op de Florida Keys, zoals deze beschreven wordt door Voss & Voss, kan worden verklaard uit locale factoren.

> Voss, G. L. & Voss, N. A., Bull. Mar. Sci. Gulf Caribb. 5, 203–229, 1955

II

Bij de systematische indeling van het geslacht *Jania* (Rhodophyta) is de grote morfologische variabiliteit binnen de soorten te weinig in het oog gehouden.

III

Mariene wieren zijn taxonomisch nog onvoldoende bekend om juiste conclusies te kunnen trekken over de geografische verspreiding der soorten.

IV

De plant bezit geen speciale inrichtingen voor het waarnemen van de richting van de zwaartekracht.

> PICKARD, B. G. & THIMANN, K. V., J. Gen. Physiol. 49, 1065–1086, 1966

V

De opvatting van DARLINGTON over de relatie tussen cytogenetica er klassieke systematiek is zeer eenzijdig.

> DARLINGTON, C. D., Chromosome botany. Allen & Unwin, Ltd. 1963

VI

De mening van TAUBER, dat de val van de *Ulmus*-curve in NW-Europese pollendiagrammen te verklaren zou zijn door uitfiltering, is slechts in een aantal speciale gevallen juist.

> TAUBER, H., Danmarks Geol. Undersøgelse, II, 89, 1–69

VII

De aanwezigheid van zure mucopolysacchariden in de follikelholte van rijpende oocyten bij *Limnaea stagnalis* verdient aandacht bij de bestudering van het ovulatiemechanisme.

UBBELS, G. A., A cytochemical study of oogenesis in the pond snail Limnaea stagnalis. Dissertatie Utrecht, 1968

VIII

De opvatting van SNYDER, c.s. betreffende de systematische indeling van het geslacht *Fusarium* verdient, vanuit phytopathologisch oogpunt, de voorkeur boven het concept dat ten grondslag ligt aan de door WOLLENWEBER gegeven indeling.

SNYDER, W. C. & TOUSSOUN, T. A., Phytopathology 55, 833—837, 1965

IX

Het verdient aanbeveling om bij het herbariummateriaal van wieren ook tekeningen te voegen van de microscopische preparaten die ten behoeve van de determinatie werden vervaardigd.

X

De oprichting van kleuterdagverblijven voor geestelijk gehandicapte kinderen dient te worden gestimuleerd. Er bestaat een duidelijke behoefte aan een gerichte opleiding voor leerkrachten aan deze instellingen.

M. VROMAN — 4 november 1968.

VOORWOORD

Bij het afsluiten van een belangrijke periode uit mijn leven wil ik gaarne dank betuigen aan allen die hebben bijgedragen om dit punt te bereiken.

Allereerst wil ik daarbij denken aan mijn ouders, die mij in staat stelden met de studie in de biologie te beginnen. Helaas heeft mijn moeder deze dag niet meer mogen beleven.

Verder wil ik mijn dank betuigen aan de hoogleraren, lectoren en leden van de wetenschappelijke staf van de Faculteit der Wiskunde en Natuurwetenschappen te Utrecht. Ik beschouw het als een voorrecht van hen onderricht te hebben ontvangen. Met eerbied herdenk ik hier de reeds overleden hoogleraren prof. dr. Joh. Westerdijk, prof. dr. A. A. Pulle en prof. dr. V. J. Koningsberger.

Hooggeleerde Lanjouw, hooggeachte promotor, Uw medewerking toen ik de wens te kennen gaf me graag in algologische richting te willen specialiseren, de grote mate van vrijheid die U mij gelaten hebt bij het bewerken van dit proefschrift en het vertrouwen dat U mij steeds geschonken hebt, zijn door mij in hoge mate op prijs gesteld.

Het Bestuur van de Stichting Wetenschappelijk Onderzoek in Suriname en de Nederlandse Antillen ben ik veel dank verschuldigd voor de geschonken mogelijkheid om gedurende bijna een jaar in de Nederlandse Antillen mariene wieren te verzamelen.

De Landsregering van de Nederlandse Antillen wil ik dank zeggen voor het toekennen van een subsidie aan de Stichting Natuurwetenschappelijke Studiekring voor Suriname en de Nederlandse Antillen waardoor het mogelijk werd deze dissertatie op te nemen in de reeks van Uitgaven van de Studiekring. Dr. P. Wagenaar Hummelinck en dr. J. H. Westermann hebben zich erg veel moeite getroost bij het persklaar maken van het manuscript.

Het College van Directeuren der Vrije Universiteit te Amsterdam ben ik zeer erkentelijk voor de toekenning van een jaar studieverlof, de hoogleraar-directeur van het Botanisch Laboratorium, prof. dr. L. Algera, voor de vrijheid bij het bewerken van de collecties.

Tijdens mijn verblijf op de Nederlandse Antillen had ik het voorrecht als een der eersten gebruik te kunnen maken van de faciliteiten van het Caraïbisch Marien Biologisch Instituut op Curaçao. Veel hulp en medewerking heb ik daar ondervonden van de toenmalige directeur, dr. J. S. Zaneveld.

Op de Nederlandse Antillen werd veel medewerking ondervonden van

de plaatselijke autoriteiten, zowel wat de huisvesting als wat de vervoers-
mogelijkheden aangaat. Omdat dit proefschrift handelt over de Bovenwindse Eilanden wil ik hier met name noemen de heer H. A. Hesling,
destijds gezaghebber van deze eilanden, en voorts de heren Buncamper
en van Delden, respectievelijk administrateur op Saba en St. Eustatius.

Bij het bewerken van het materiaal heb ik medewerking gekregen van
verschillende doctoraal-studenten in de biologie aan de Vrije Universiteit
te Amsterdam. Graag wil ik ook hen bedanken. Het is prettig dat enkelen
van hen zich ook na het doctoraal-examen met de studie der wieren
bezighouden.

Bij het bestuderen der collecties heb ik meermalen een dankbaar
gebruik mogen maken van de gastvrijheid van het Rijksherbarium te
Leiden. De hulp en de belangstelling, die ik daarbij steeds mocht onder-
vinden van mej. dr. J. Th. Koster wordt op hoge prijs gesteld. Zij bewerkte
ook het materiaal van de Cyanophyta.

Het was een voorrecht, dat ik ook in de gelegenheid was de voor de
studie van de algen uit het West-Indische gebied zeer belangrijke collecties van M. A. Howe in het Smithsonian Institution in Washington (D.C.)
en van F. Børgesen in het Institut for Sporeplanter te Kopenhagen te
bestuderen.

Veel belangstelling en steun kreeg ik ook van prof. dr. W. R. Taylor
te Ann Arbor (Michigan), die zelf een groot deel van zijn leven wijdde
aan de studie van de algen van het Westelijke, tropische deel van de
Atlantische Oceaan.

Het tekenwerk werd verzorgd door de heer G. W. H. van den Berg,
het fotowerk door de heer C. van Groeningen. Beiden wil ik bedanken
voor de keurige uitvoering.

Drukkerij Kemink en Zn. te Utrecht verzorgde het drukwerk op een
zeer prettige en accurate wijze.

Wanneer wieren gedetermineerd zijn volgt nog een hele reeks van
handelingen voor ze goed en wel op hun plaats in het herbarium zijn
beland. Allen die daarbij hun hulp verleenden wil ik ook heel hartelijk
bedanken voor de zorg waarmee ze dit werk hebben gedaan. De belang-
rijkheid ervan wordt te vaak onderschat.

CONTENTS

1

LIST OF FIGURES

LIST OF PHOTOGRAPHS

4

CHAPTER I

INTRODUCTION

Although algology deals with a large group of plants, widespread and of a great morphological diversity, the history of this branch of botany is fairly young.

LINNAEUS (1753) listed in his *Species Plantarum* under the heading *"Cryptogamia — Algae"* only five genera of plants which are still accepted as algae at the present time. Under the same heading he also described a number of liverworts, lichens and sponges and a few other things.

The five genera of algae described by LINNAEUS are:

C h a r a — stoneworts;

F u c u s — brown algae; next to several *Fucus*-species which are still valid, also species belonging to the genera *Ascophyllum*, *Sargassum*, and *Turbinaria*;

U l v a — green algae with a flat or hollow thallus, including species of the present genera *Ulva* and *Enteromorpha*;

C o n f e r v a — filamentous green algae, inter alia species of the genus *Cladophora*;

B y s s u s — thin, threadlike floating algae.

That LINNAEUS was a good observer is clear from the fact that most of his species are still recognized. Moreover, his herbarium contains specimens of many of the larger forms.

After LINNAEUS, innumerable new algal species have been described. The period of scattered descriptions of new forms ended with the appearance of the first local floras and with C. A. AGARDH's general encyclopedical work *Species algarum rite cognitae* (1821—1828). After C. A. AGARDH two other authors have tried to present a complete algological knowledge: J. G. AGARDH in *Species, genera et ordines algarum* (1848—1901), J. B. DE TONI in *Sylloge algarum* (1889—1924).

During the last few decades, however, algological knowledge has increased to such an extent that it must be considered impossible to cover the whole field of phycology in one general work. Fortunately, from the nineteenth and twentieth centuries we have many local algal floras, a survey of which has been published by TAYLOR (1959b).

The history of algology, especially for the West Indies, has been treated at length by TAYLOR (1960) in his large and comprehensive work *Marine algae of the eastern tropical and subtropical coasts of the Americas*. After a stay at the Dry Tortugas Laboratory of the Carnegie Institution of Washington during the summer months of 1926—1928, TAYLOR published many articles and papers on marine algae from most parts of the Caribbean. Nevertheless only a few places have been thoroughly investigated.

As our investigation concerns the algal flora of the Windward Group of the Lesser Antilles (see map), the most important algological publications for this region will be mentioned.

An important work is BØRGESEN's (1913—1920) thorough and detailed study of the marine algae of the Virgin Islands (then Danish West Indies).

Also well-known is the work on algae of Guadeloupe, collected by MAZÉ & SCHRAMM (1870—1877) and identified by the brothers CROUAN at Brest. The material of MAZÉ & SCHRAMM was collected very carefully; however, the brothers CROUAN did not realize the enormous variability of marine algae. Unfortunately, a complete set of this collection has never been deposited in any of the larger herbaria of the world, although several institutes possess large portions of MAZÉ & SCHRAMM's collections. A closer study of the work of MAZÉ & SCHRAMM reveals that many of their newly described names are *nomina nuda*, and their publication should be regarded with great caution.

HAMEL & HAMEL-JOUKOV (1929—1931) and QUESTEL (1951) published supplementary data on Guadeloupe.

The marine algae of Barbados have been studied by Miss VICKERS. In 1905 she published a fairly short list of species, followed in 1908 by an atlas with excellent figures for a number of *Chlorophyceae* and *Phaeophyceae*.

The Bahamas, north of the Lesser Antilles, are well known by the work of HOWE (1920).

In a number of publications TAYLOR (1929, 1940, 1942) presented a survey of the algal flora of several islands of the Lesser Antilles. The results of these investigations have also been incorporated in his algal flora for the West Indies (1960). In 1962 his supplementary paper on the *Algae from the Lesser Antilles* appeared. He was able to study a fairly large number of collections from the British Virgin Islands, Barbuda, Nevis, Antigua, St. Kitts, St. Lucia and Grenada. Nevertheless, TAYLOR (1962, p. 44) stated: "No general conclusions can be derived from these collections — they fill in gaps in our knowledge, but do not change our general picture of the West Indian algal flora".

As for the Netherlands Antilles, not much can be reported. Marine algae have been collected at several instances, but only a small part of the collections has been studied.

SURINGAR, one of the leaders of the "Nederlandsche West-Indische Wetenschappelijke Expeditie 1884/85" collected a number of marine algae on these islands.

In 1905 J. BOEKE, a fishery expert, also sampled seaweeds.

6

TAYLOR (1939) paid two short visits to Aruba and Curaçao and took several samples of marine algae on that occasion.

WAGENAAR HUMMELINCK has made large collections of marine organisms during several trips to the West Indies (1930, 1936/37, 1948/49, 1955, 1963/64, 1967).

In 1954/55 ZANEVELD collected seaweeds of Curaçao, Aruba and Bonaire.

Financial aid of the Netherlands Foundation for the Advancement of Research in Surinam and the Netherlands Antilles (WOSUNA) enabled the author to collect marine algae in the Netherlands Antilles, with the Caribbean Marine Biological Institute as an operational base, from November 1957 till July 1958. Particularly along the coasts of Curaçao many specimens were collected. Special trips were made to Aruba (April 3—15), Bonaire (Feb. 20—22, March 17—29), and the Windward Group (April 24—May 31). From St. Martin short visits were made to St. Eustatius (May 19—22) and Saba (May 27—30).

At the same time DÍAZ-PIFERRER visited Curaçao (1958), and a few years later also Bonaire (1962).

Only a small part of these collections has been studied.

Miss SLUITER (1908), under the guidance of Mrs. WEBER-VAN BOSSE, studied the collection of BOEKE, together with a few specimens collected by Naval-Commander VAN SCHOONHOVEN. A short list of names and collecting places was published and about 60 species recorded.

The bluegreen algae of Aruba, Bonaire and Curaçao, collected in 1930 by WAGENAAR HUMMELINCK, were identified by FRÉMY (1941).

TAYLOR (1942) reported on his material from Aruba and Curaçao in *Caribbean marine algae of the Allan Hancock Expedition, 1939.*

Miss KOSTER (1943, 1960) studied a number of samples of green and bluegreen algae collected by WAGENAAR HUMMELINCK from fresh and brackish waters in the Netherlands Antilles. She also investigated (1963) samples of marine Cyanophyceae of the Lesser Antilles from the collections of SURINGAR, WAGENAAR HUMMELINCK and VROMAN, and some material from salt pans on Bonaire and Curaçao.

The list of publications on the marine algae of the Netherlands Antilles is closed by that of DÍAZ-PIFERRER (1964), which gives a number of supplementary data on the algal floras of Curaçao and Bonaire, based on his own collections.

NATURE AND LIMITATION OF THE INVESTIGATION

The study of the extensive collection of marine algae collected in the Netherlands Antilles during our stay in 1957/58 was started with the material from the islands of St. Martin, St. Eustatius and Saba (Lesser Antilles). The algal vegetation of Aruba, Bonaire and Curaçao will be treated in separate publications.

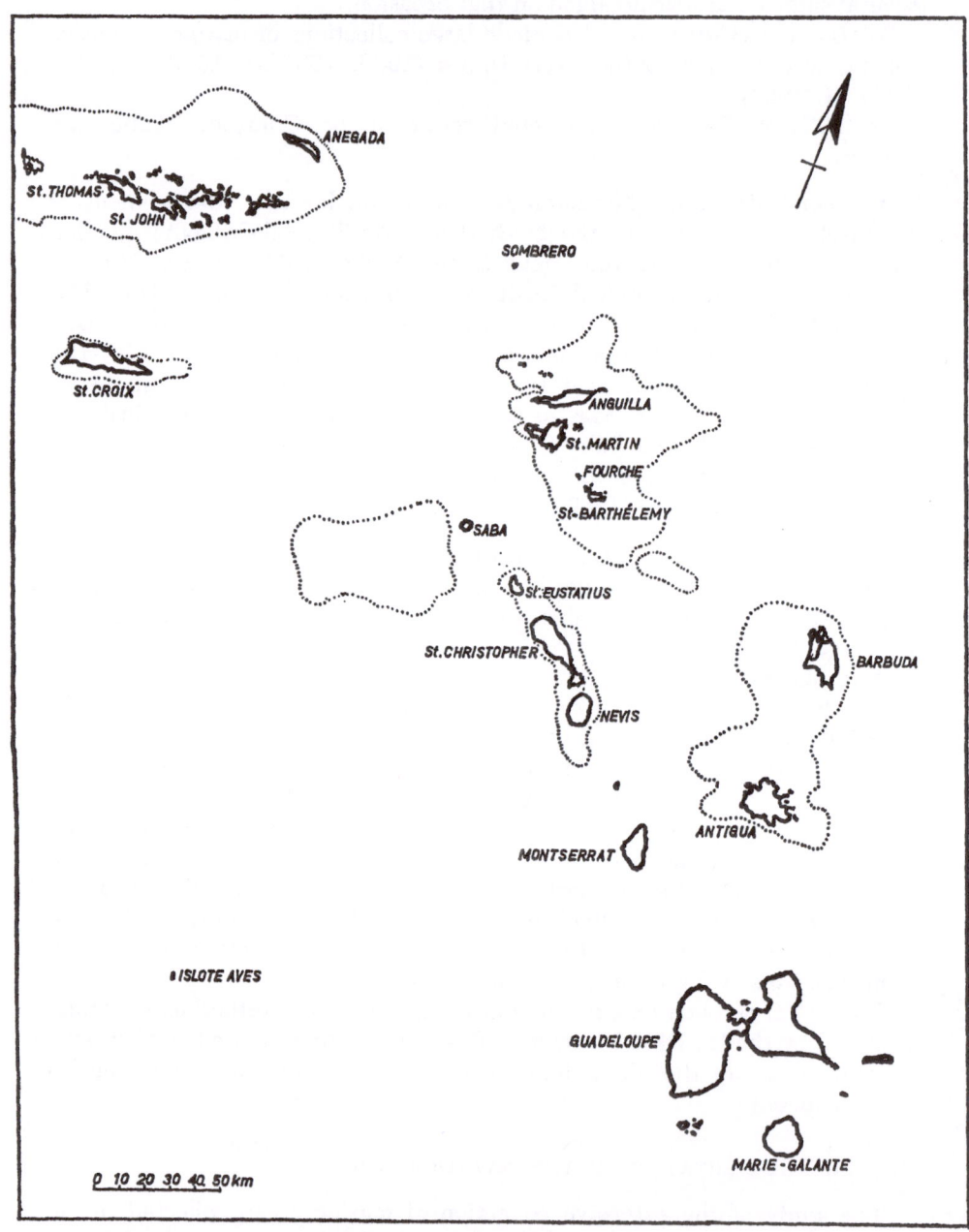

Fig. 1. Sketch map of part of the WINDWARD GROUP of the Lesser Antilles.

On St. Martin many samples were taken. The coastal configuration of the island varies highly, and therefore exact descriptions were made of the algal vegetation in different places. The material was collected by walking along the shore or by swimming and diving, and therefore it only contains forms growing at a depth of 5 meters or less. In a few cases material was collected that was washed ashore, and possibly originated from greater depths.

This treatise contains a short outline of the geology, the coastal forms, and the habitat factors mainly based on data from the literature (Chapter II), and a survey of the algal vegetations in relation to coastal forms (Chapters III & IV).

The material collected on these islands by SURINGAR in 1885 (deposited in the Rijksherbarium in Leiden) was also studied, as was the material collected by WAGENAAR HUMMELINCK on St. Martin, St. Eustatius and Saba in 1949, supplemented with material from St. Martin, gathered in 1955 — together with a number of samples which were taken by the same collector on St. Thomas, St. John, St. Croix, Anguilla, Fourche, St. Barts, St. Kitts, Nevis, Barbuda and Islote Aves in 1937, 1949 and 1955. The results of this study have been summarized in Chapters V and VI which also contain an alphabetical list of species, and a few remarks on nomenclature and systematic position.

Chapter VII gives an enumeration of all species collected, listed per collecting place.

The material of SURINGAR will be reviewed in a separate publication.

Attention is drawn to the following topographical names, which are generally used in different meanings; in this paper they cover the areas as described:

West Indies	Antilles, Bahamas, Florida Keys, Bermuda, Cayman Islands, Swan Island, Old Providence, San Andrés
Antilles	Cuba to Trinidad and Aruba
Greater Antilles	Cuba to Puerto Rico
Lesser Antilles	Virgin Islands to Trinidad and Aruba
Windward Group . .	Virgin Islands to Grenada (Bovenwindse Eilanden, Islas de Barlovento, Iles sur le Vent, Inseln über dem Winde)
Leeward Islands . .	(British usage) Virgin Islands to Dominica
Windward Islands .	(British usage) Martinique to Grenada
Leeward Group . . .	Los Testigos to Aruba and Los Monges (Benedenwindse Eilanden, Islas de Sotavento, Iles sous le Vent, Inseln unter dem Winde)

ACKNOWLEDGEMENTS

The author wishes to express his gratitude to:
the Netherlands Foundation for the Advancement of Research in Surinam and the Netherlands Antilles (WOSUNA) for the financial aid which enabled him to collect marine algae in the Caribbean;

the Netherlands Antilles Government for granting a subsidy to the Foundation for Scientific Research in Surinam and the Netherlands Antilles (Studiekring) to cover the costs of publication of this treatise;
the 'College van Directeuren' of the Free University at Amsterdam for a one-year's leave to study the algae of the West Indies and to the director of the Botanical Laboratory, prof. dr. L. ALGERA, for the possibilities given afterwards to work on the extensive collections.

Thanks are also due to:
prof. dr. J. LANJOUW, dr. P. WAGENAAR HUMMELINCK and dr. J. H. WESTERMANN for their useful criticism and help;
dr. J. S. ZANEVELD, former director of the Caribbean Marine Biological Institute at Curaçao, for his assistance;
the local authorities of the Government for rendering much help with regard to housing and transport on the islands; especially to be mentioned are mr. H. A. HESLING, former governor of the Windward Group, and messrs. VAN DELDEN and BUNCAMPER, former administrators of St. Eustatius and of Saba.
The drawings were made by mr. G. W. H. VAN DEN BERG.

Finally the author would like to acknowledge his great indebtness to the Foundation's Secretariat, for its great help and expert assistance in preparing this treatise for the press.

CHAPTER II

HABITAT FACTORS

A. GEOLOGY AND COASTAL FORMS

The geology of the islands has been described in detail by CHRISTMAN (1953; St. Barts, St. Martin, and Anguilla) and WESTERMANN & KIEL (1961; Saba and St. Eustatius).

ST. MARTIN

St. Martin (Fig. 2) is situated on the outer arc of the Lesser Antilles, which is characterized by the absence of recent or geologically young volcanoes.

Its geological formations are shown in the sketch-map. There is a close relationship between the formations exposed on the coast and the morphology of the coast. Sandy beaches and slightly curved sand bars or spits, separating lagoons from the sea, alternate with rocky coasts formed by limestones and marls (belonging to the Quaternary and Low Lands formation) or by extremely hard Point Blanche sedimentary rocks, dolerites (basalts) and porphyrites.

In Great Bay a sand bar at a depth of $2\frac{1}{2}$—$3\frac{1}{2}$ metres roughly parallels the sand bar on which Philipsburg is situated; it is clearly visible from above since it lacks the growth of seagrasses.

The sand beaches and bars contain few coral debris notwithstanding the rich coral growth offshore. In places the sand and debris have been cemented into so-called beachrock (Pl. IXa), which must be very young; it contains shells which still have their natural colour.

Cliff coasts of Quaternary coral limestone — somewhat similar to those described by DE BUISONJÉ & ZONNEVELD (1960) from Aruba, Curaçao and Bonaire — are found in Little Bay (Pl. IIIb), Point Blanche Bay (Pl. IVb), Guana Bay (Pl. IVa), Oyster Pond and other places. The limestone generally shows a "lapiés" or "Karren" habit. The cliff of Point Blanche Bay has a clearly discernible niche, below which a horizontal bench, which is constantly washed by the waves, projects.

The limestone and marl cliffs of the Low Lands formation are in places bordered by a sandy beach (Pl. II, Mary Point). The coastal water is shallow and the sandy bottom is covered with extensive fields of the seagrasses *Syringodium filiforme* and *Thalassia testudinum*. Where the steep cliff, which is 7—10 m high, rises directly from the sea, algal growth is absent due to the action of sand-loaded waves.

Between Cole Bay and Long Bay the bays are separated by Low Lands

11

formation rocks projecting into the sea, reaching a few metres above sea-level. The rocks show sharp ridges and edges, and in places niches are formed by wave action (Pls. Ia, V, VI).

Hard Point Blanche sedimentary rocks exposed on the coast are shown in Plates VII—VIII. Usually the beach is covered with many large, partly

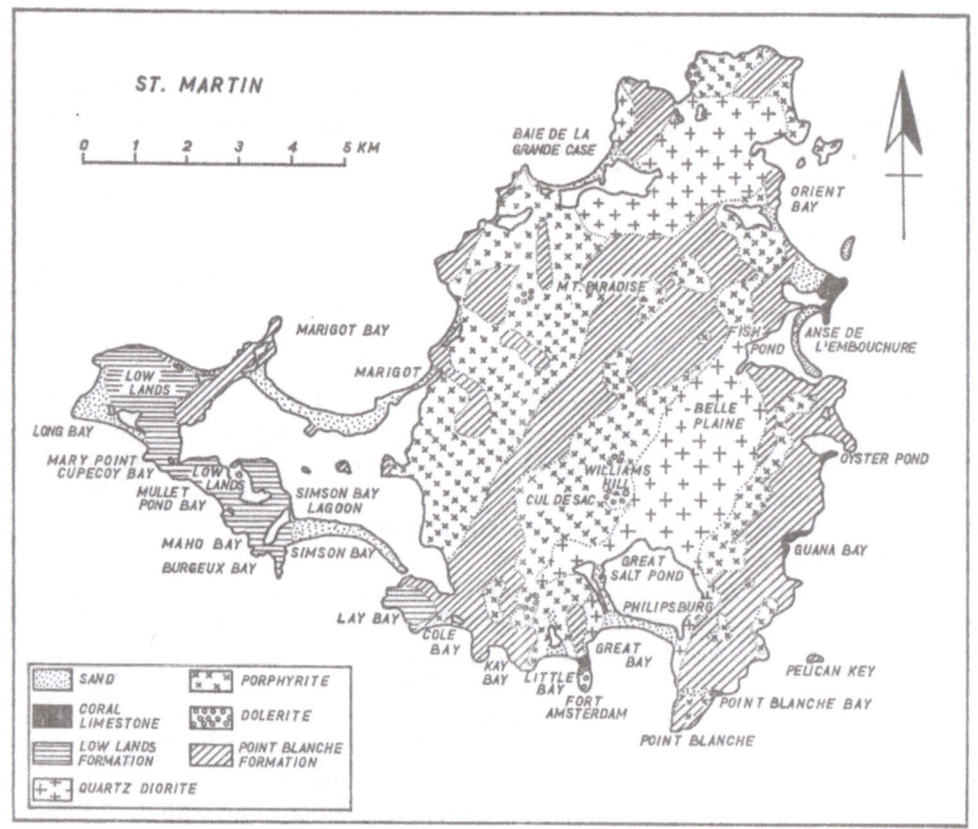

Fig. 2. Geological sketch map of ST. MARTIN (after CHRISTMAN, 1953).

smooth-surfaced boulders, the diameter of which may be more than 1 metre. Marine organisms can easily attach themselves to the furrowed sides of these stratified boulders.

At the peninsula of Fort Amsterdam (Pl. IIIa) the coast is formed by hard dolerites (basalts), rising almost vertically from the sea, and beaches of rounded and polished boulders.

In Baie de la Grande Case (Pl. IXa) part of the coast is formed by large blocks of quartz diorite with diameters of 50 to 200 cm; algae can easily attach themselves to their rough surface.

Saba (Fig. 3) forms part of the inner or volcanic Lesser Antilles arc, characterized by the occurrence of recent or young volcanoes.

The extinct Saba volcano rises steeply from a depth of about 650 m. It has no crater and consists of lava domes, lava flows, tuffaceous and agglomeratic beds, largely of an andesitic nature. Convex coastal lines are

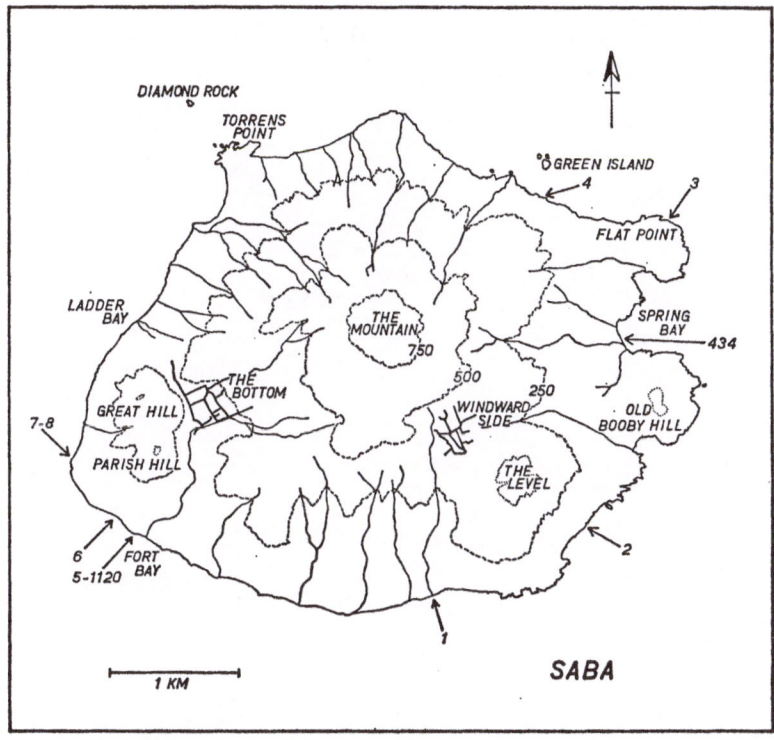

Fig. 3.   Sketch map of SABA, with collecting numbers
(after KLM-Aerocarto 1959, 1 : 20,000).

formed by the hard rocks of the peripheral lava domes and locally by lava flows such as that of Flat Point; in these places the coast is steep and inaccessible, with many large blocks. Concave coastal lines are found where less coherent tuffs and agglomerates offer relatively little resistance to marine erosion, resulting in pebbly beaches as in Spring Bay and Ladder Bay. A sandy bottom is found in Fort Bay.

St. Eustatius

St. Eustatius (Fig. 4) is also one of the islands of the inner or volcanic Lesser Antilles arc. It consists of a greatly denuded, northwestern andesite volcano, and a young, extinct ash volcano in the southeast, the Quill. The latter is beautifully cone-shaped and has a wide and deep crater. Tilted slabs, consisting largely of limestone, rest against the south slope of the Quill volcano (White Wall, Sugar Loaf; Pl. X).

Where the Quill volcano borders the sea, a steep cliff has been formed showing horizontal stratification of agglomerates and tuffs. The beach

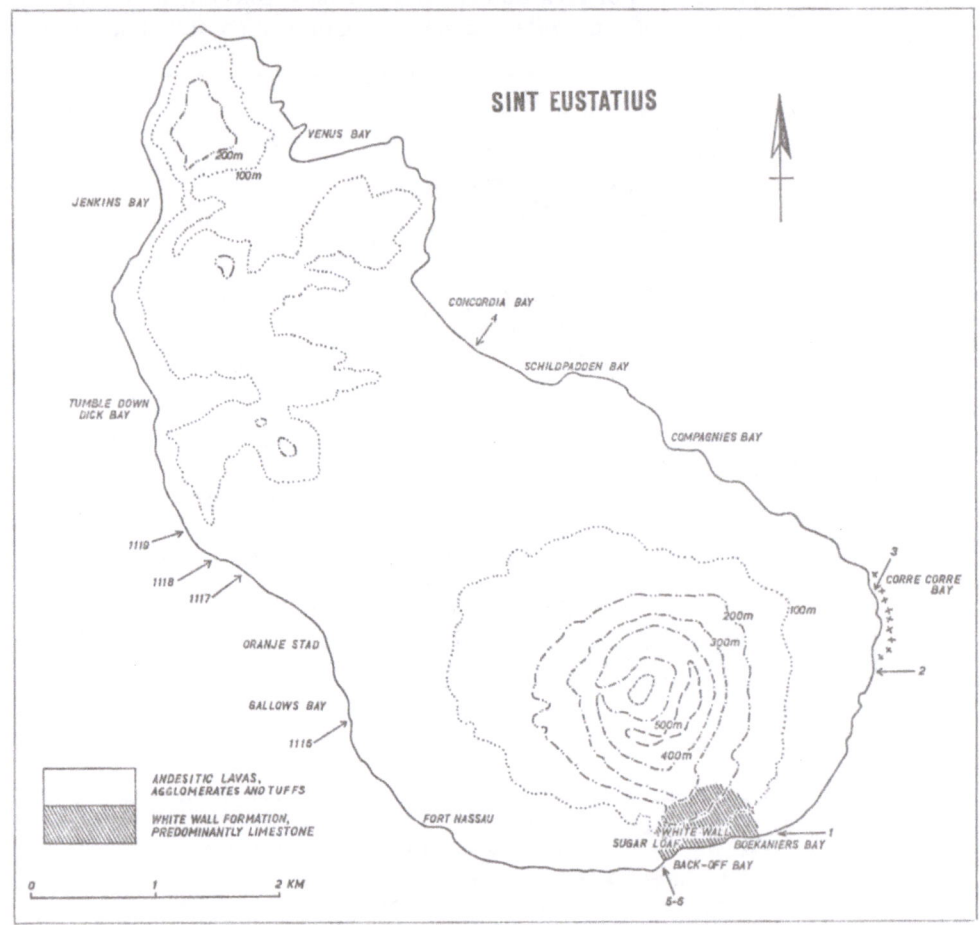

Fig. 4. Sketch map St. Eustatius, with collecting numbers (after Wagenaar Hummelinck, 1953).

below the cliff consists largely of pebbles. A beautiful coral reef is present at Corre Corre Bay, about 75 metres offshore; between the coast and the reef the water is fairly quiet and offers excellent conditions for algal growth. Elsewhere wide sandy beaches occur, locally covered with pebbles and boulders; the sand may be of a dark colour owing to the abundance of magnetite and ilmenite.

White Wall and Sugar Loaf descend steeply into the sea; nearby is a beach with large boulders (Pl. X).

14

B. Climate

A comprehensive survey up to 1933 of the meteorological observations in Surinam and the Netherlands Antilles is given by BRAAK (1935). From 1953 onwards annual meteorological data for the Netherlands Antilles are being published by the Department "Economische Zaken en Welvaarts-zorg, Bureau voor de Statistiek", Curaçao. From 1959 onwards also detailed observations are given for the Lesser Antilles, mostly for St. Martin.

Wind

The trade winds blow throughout the year from directions between ENE and E. The Windward Group of the Lesser Antilles is situated in the West Indian hurricane zone. The passage of hurricanes is generally accompanied by abnormally heavy rains.

BRAAK (1935) gives the following data for Philipsburg (St. Martin), and Oranjestad (St. Eustatius) (Table 1). The velocity is greatest during the

Table 1

WIND VELOCITY AND WIND DIRECTION IN ST. MARTIN AND ST. EUSTATIUS

	St. Martin			St. Eustatius		
	average wind velocity (in m/sec)	mean direction of wind (N → E)	steadiness (in %)	average wind velocity (in m/sec)	mean direction of wind (N → E)	steadiness (in %)
Jan.	3.9	73°	91	5.3	63°	87
Feb.	3.3	78°	90	4.9	63°	83
Mar.	3.3	80°	84	4.5	68°	87
Apr.	3.5	81°	92	4.9	67°	83
May	3.1	85°	91	4.7	80°	81
June	3.7	82°	95	5.1	80°	87
July	4.1	74°	96	5.3	78°	88
Aug.	3.7	77°	92	4.9	73°	86
Sep.	3.3	78°	88	4.1	82°	80
Oct.	3.0	79°	90	3.3	77°	80
Nov.	3.3	76°	88	3.9	71°	84
Dec.	4.1	72°	95	4.3	70°	87

Table 2

MEAN WIND VELOCITY IN ST. MARTIN AND ST. EUSTATIUS (in knots)

local time	St. Martin	St. Eustatius
8.00 h	9.6	5.3
14.00 h	11.4	6.2
20.00 h	7.9	4.4

middle of the day, as is shown in Table 2 for Juliana Airport (St. Martin) and for Oranjestad (St. Eustatius) copied from the Statistics for the year 1959. The difference in the wind velocity for both islands, not reported by BRAAK, may be explained by the location of the observation stations.

AIR TEMPERATURE

The air temperatures show little fluctuation. The difference between the highest and lowest monthly mean temperature is only 3° C. On the whole temperatures are very tolerable on account of the eastern trade wind.

BRAAK (1935) gives the following figures for Oranjestad (St. Eustatius) (Table 3).

Table 3

AIR TEMPERATURE IN ORANJESTAD, ST. EUSTATIUS (mean temperature in °C)

1910—1918

Jan.	24.2		July	26.6
Feb.	24.0		Aug.	27.0
Mar.	24.2		Sep.	27.0
Apr.	24.8		Oct.	26.6
May	25.7		Nov.	26.1
June	26.4		Dec.	25.1
		Yearly average 25.6		

RAINFALL

The rainfall on the Lesser Antilles measures about 1000 mm a year. The rain mostly comes down in short, heavy downpours; immediately afterwards the sky brightens again. Tables 4 and 5 present the rainfall in mm and also the number of days with 1.0 mm or more rain.

The heavy rainfall during the first 5 days of May 1958 caused a strong soil erosion. The water of Great Bay was brown-coloured on May 6 owing to the silt; this was especially observed in the northwestern part of Great Bay where Freshwater Pond discharges.

16

Table 4

RAINFALL IN ST. MARTIN, ST. EUSTATIUS AND SABA

(in mm; *Statistics meteorol. observ. Neth. Antilles* 6, 1958, table I)

Rainstation	1958													1956	1957
	Jan.	Febr.	Mar.	Apr.	May	June	July	Aug.	Sep.	Oct.	Nov.	Dec.	Year	Year	Year
St. Martin															
Juliana Airport	10.2	30.2	16.0	92.7	269.5	86.3	119.4	111.7	105.4	193.6	164.3	58.9	1258.2	1361.0	828.8
Philipsburg	6.7	38.3	45.3	14.0	308.5	38.2	106.8	93.7	68.3	224.2	100.7	59.9	1104.6	1560.0	?
St. Eustatius															
Oranjestad	7.4	20.8	12.7	18.8	316.0	149.6	186.6	96.0	216.6	200.8	105.6	57.1	1388.0	?	909.1
Saba															
The Bottom	18.8	40.0	2.0	40.5	352.1	88.2	37.9	11.0	133.5	231.0	151.0	69.7	1175.7	1167.7	1128.1
Windwardside	31.3	18.7	2.0	49.4	383.0	86.4	101.4	108.7	85.9	218.3	98.5	60.6	1244.2	1074.5	990.2

Table 5

Number of days with 1 or more mm precipitation in St. Martin, St. Eustatius and Saba

(*Statistics meteorol. observ. Neth. Antilles 6, 1958, table II*)

Rainstation	1958													1957	1956
	Jan.	Febr.	Mar.	Apr.	May	June	July	Aug.	Sep.	Oct.	Nov.	Dec.	Year	Year	Year
St. Martin															
Juliana Airport	4	7	2	8	14	15	20	15	17	15	17	13	147	158	145
Philipsburg	1	6	2	4	12	5	8	8	10	17	11	14	98	?	130
St. Eustatius															
Oranjestad	2	5	4	7	12	19	19	8	15	7	14	12	124	129	?
Saba															
The Bottom	4	8	1	7	11	14	10	5	15	13	14	19	121	128	110
Windwardside	6	4	1	7	13	10	13	6	10	17	13	12	112	123	143

C. WATER MOVEMENTS

SEA CURRENTS

The sea currents around St. Martin, Saba and St. Eustatius are little known but are no doubt influenced by the movements of the Antillean current along the outer side of the Lesser Antilles, and the Gulfstream. The mean annual temperature of the water is 25° C; even during the coldest month the temperature is not below 20° C (EKMAN, 1953).

TIDAL MOVEMENTS

For St. Martin, St. Eustatius and Saba no direct information on the tidal movements is available.

On Puerto Rico a 12-hour rhythm has been observed (BIEBL, 1962, who has used the figures from COKER & GONZALEZ, 1960). During the month of June low water is at noon, in December at midnight. The tidal differences are only 10—30 cm.

On Barbados the tides are of a mixed biurnal type, with high and low water twice a day (LEWIS, 1960, see Fig. 5; cf. also WAGENAAR HUMMELINCK, 1953, fig. 1, and DE HAAN & ZANEVELD, 1959, fig. 2). The mean tidal difference on Barbados is about 2.3 feet (70 cm), the maximum 3.6 feet (110 cm).

In large areas of the Caribbean, including St. Martin, Saba and St. Eustatius, the tidal movements are a combination of the Puerto Rican and Barbados' types.

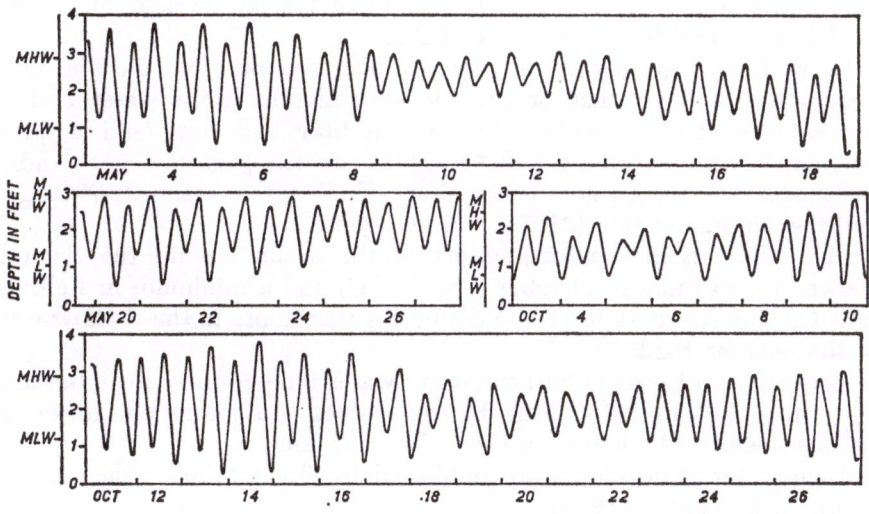

Fig. 5. Tide curves for BARBADOS in May and October 1958
(after LEWIS, 1960).

19

Data on the mean sea level throughout the year are published by PATULLO et al. (1955) and DE HAAN & ZANEVELD (1959).

The period of submergence of rocks in the tidal zone during the various seasons is of great importance for organisms living in the tidal zone. DE HAAN calculated the period of submergence for different levels on Curaçao. His "standard submergence table" applies only to this island and is based on a limited period of observation (DE HAAN & ZANEVELD, 1959, fig. 7 and table 4). HUMM (1963) described this yearly fluctuation of sea-level for the southern Gulf of Mexico.

During periods of low sea-level in the summer months most of the algae in higher places die; this was observed by the present author in several places.

WAVE ACTION

The eastern shore of the islands is exposed to heavy surf (Pls. Ia, IIIa, IV, VIIa, VIIIa, IX, X). The western shore and generally also the southern and northern coasts are much less exposed (Ib, II, IIIb, Vb, VIb, VIIb, VIIIb, IXa). On days with strong winds, wave action on the eastern shore is 8 times stronger than on the western shore (LEWIS, 1960; Barbados).

The niches in the coral limestone of strongly exposed coasts are much wider than those in sheltered bays (DE BUISONJÉ & ZONNEVELD, 1960; Curaçao).

D. PROPERTIES OF SEA-WATER

SURFACE TEMPERATURE

Exact data on the surface temperature of the sea around St. Martin, St. Eustatius and Saba are not available.

SMITH (1940) gives data on Great Bahama Bank (Fig. 6). The water temperature follows that of the air very closely. Both water and air temperature show a marked increase in May and June and a strong decrease between October and November. In the Middle Bight, Andros, the maximum mean water temperature (29.7° C) is reached in August, and the minimum value (21.6° C) in December.

Measurements of the temperature of the ocean around the Bahamas showed a maximum in October (28—29° C) and a minimum in February (23° C). The fluctuations are no doubt less than those in the shallow water of the Bahama Bank.

According to LEWIS (1960) the mean water temperatures near Barbados vary from 25° to 28° C; the highest values are measured during the summer months, the lowest in December and January.

Figures for Puerto Rico are published by BIEBL (1962) who refers to COKER & GONZALEZ (1960).

In very shallow coastal waters, for instance in bays and lagoons, the water temperature may show proportionately large differences.

20

SALT CONTENT

Much information on the composition of sea-water of the Caribbean Sea and Cayman Sea is found in RAKESTRAW & SMITH (1937), PARR (1937, 1938), and in papers on local investigations, for instance SMITH (1940), LEWIS (1960) and BIEBL (1962).

The Cl-content of the surface water of the open sea surrounding the Lesser Antilles is about 36.2—36.4⁰/₀₀. Near the coast there may be considerable fluctuations in the salt content; this is particularly the case in lagoons almost completely separated from the sea. During long periods of drought there is a strong evaporation of the water in the lagoons, and even crystallisation of salt may then occur. On the other hand, during long periods of rainfall the water may become less salty and sometimes brackish; at the same time a great deal of silt may be carried into the lagoon. Only a few organisms are adapted to these extreme environmental fluctuations.

SILT CONTENT

Periodically the silt content of the water near the shore may be very high. Usually the silt will settle down after some time, especially in the back part of bays and lagoons. *Avicennia germinans* (= *A. nitida*) is practically the only plant able to maintain itself in the black and oxygen-deficient bottoms; the pneumatophores are nearly always devoid of algae.

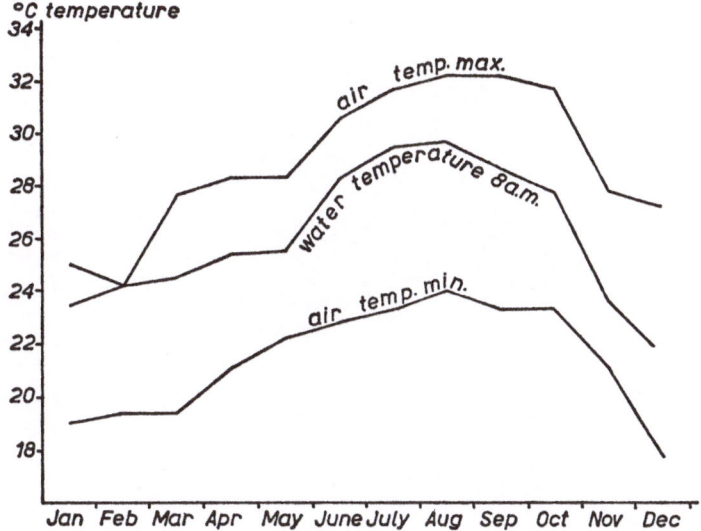

Fig. 6.   Monthly mean air and sea temperatures based on daily records at the Nassau Meteorological Station, BAHAMAS. Sea temperatures measured in Middle Bight, Andros, at 8 A.M. (SMITH, 1940).

E. SUBSTRATE

The coasts of the Lesser Antilles are formed by rocks of different origin (see geological sketch-maps).

It is wellknown that algae find much better possibilities for attachment on one kind of rock than on another. On the roots of *Rhizophora mangle*, algal species also find favourable conditions for attachment and shelter against strong insolation in the shade of the foliage. These species need a firm basis for attachment and therefore can also be found on rocky coasts. Finally several tropical marine algae are able to establish and maintain themselves in loose sandy and muddy bottoms.

F. BIOTIC FACTORS

Man-made pollution of the habitat is uncommon on the Lesser Antilles. Near the centers of population some pollution by sewage may result in a dense growth of green algae. Oil pollution is unknown in St. Martin, St. Eustatius and Saba.

Quantitative data on the feeding of fishes and snails on marine algae are not available, and it is impossible to assess its damage to the vegetation.

In one case the contents of the stomach of an angel fish from Great Bay, St. Martin, could be analised; it contained remains of *Caulerpa sertularioides* and *Codium* spec. and also remnants of a sponge.

CHAPTER III

SURVEY OF THE LITERATURE ON THE ALGAL VEGETATIONS
IN THE CARIBBEAN

A. Introduction

Several of the algologists working in the Caribbean have given a general description of the algal vegetations of the islands.

Exact descriptions of the algal vegetations of the Danish West Indies (now U.S. Virgin Islands) were published by Børgesen (1900, 1909, 1911). As a rule tropical seaweeds are not very large; the large brown algae, so characteristic for the tidal zone of temperate regions, are absent in the tropics. As the tidal differences are small, and on the other hand sunshine is strong, hardly any algae are found above sea-level; only on rocks which are constantly washed by the waves, a number of species can be found which normally also occur in the upper part of the sublittoral region.

The algae of rocky coasts are defined by Børgesen as "lithophilous" algae, i.e. algae requiring a hard substrate.

The upper part of the sublittoral region contains inter alia *Ulva lactuca* var. *rigida*, *Enteromorpha* spp., *Struvea anastomosans*, *Neomeris annulata*, *Anadyomene stellata*, *Cladophoropsis membranacea*, and also species of the genera *Padina*, *Sargassum*, *Laurencia* and *Bostrychia*. At a somewhat lower level a great number of other species is found, for example *Caulerpa racemosa* (several varieties), *Dictyosphaeria cavernosa*, *Cladophora* spp. *Bryopsis* spp., *Codium* spp. In many places extensive vegetations of *Sargassum vulgare*, *Sargassum platycarpum* and *Turbinaria turbinata* are found, and species of the genera *Padina* and *Dictyota*, *Hydroclathrus clathratus*, *Colpomenia sinuosa*, *Bryothamnion triquetrum*, *Centroceras clavulatum*, *Ceramium nitens*, *Hypnea musciformis*, etc. Calcareous red algae of the genera *Jania*, *Amphiroa*, *Liagora* and *Galaxaura* are also abundantly present.

Much attention is given by Børgesen to the algal vegetation of lagoons and the quiet water behind coral reefs (1909, 1911).

The sandy bottom contains, between extensive patches of sea grasses, an algal vegetation rich in individuals (but not in species). These algae are indicated as "psammophilous" or "pilophilous" algae. They can be divided into two groups, the creeping algae (*Caulerpa*), and algae anchored in the bottom with a bundle of rhizoids (species of the genera *Halimeda*, *Penicillus*, *Udotea*, etc.).

On the roots of *Rhizophora mangle*, which generally forms a dense vegetation along the shore of the lagoons many "rhizophilous" algae are found. The majority of the species from the roots of mangroves appear to be also abundant in the sublittoral region of rocky coasts. Consequently, the distinction between lithophilous and rhizophilous is not sharp.

FELDMANN & LAMI visited Guadeloupe during the spring of 1936. Their publications deal with the algal vegetations of mangroves (1936) and the algae of open coasts (1937).

A distinction is made between 1) the vegetations on rocky bottom, and 2) the vegetations on sandy bottom, including mangrove-thickets.

Only a few rocky bottom species belonging to the tidal zone are mentioned (*Ralfsia expansa, Ectocarpus breviarticulatus, Chaetomorpha media, Chnoospora fastigiata* and Melobesiaceae). The sublittoral region appears to be richer in species. In exposed places *Sargassum polyceratium, S. platycarpum* and *Turbinaria turbinata* occur in the upper part of the sublittoral region, in more protected places *Padina sanctae-crucis, P. gymnospora, Colpomenia sinuosa, Hydroclathrus clathratus* and other species are found.

Unstable sandy bottoms of the tidal zone are poor in algae: only in quiet places may *Vaucheria* be found. The sublittoral region contains extensive areas of *Thalassia testudinum* and *Syringodium filiforme* *). Between these seagrasses representatives of the order Siphonales are found, e.g. *Penicillus, Rhipocephalus, Halimeda, Udotea, Avrainvillea, Caulerpa* and *Ernodesmis*. Other algae are less abundant and mostly grow as epiphytes on the leaves of *Thalassia* and on the green algae of the order Siphonales.

Most of the publications of TAYLOR deal with various small areas (1942, 1951, 1954); his book on the *Marine algae* (1960) is of importance for the whole Caribbean territory.

There are only a few publications dealing with the algal vegetations of the Netherlands Antilles. They are mentioned in the first chapter.

In the following paragraphs we will deal with a number of publications in which special attention is given to the phenomenon of zonation in the Caribbean region.

B. ZONATION IN GENERAL

One of the most remarkable features along sea coasts is the zonation of organisms in the tidal zone.

Much has been published on this subject and it appears far from easy to give a survey of the literature. Many investigators studied the problems of intertidal biology in great detail but on a small scale. General descriptions of larger stretches of coast almost do not exist; nowadays, however, it has become easier to carry out this type of investigation. An important development of the present time is also (T. A. & A. STEPHENSON, 1949) "an increasing willingness to include both the plants and the animals in marine ecological studies . . . which are in nature so inextricably interrelated that no balanced account can be achieved without both".

T. A. & A. STEPHENSON (1949) have tried to give a general survey of the common principles of zonation of the intertidal zone, aiming at a general

---

*) *Syringodium filiforme* is used instead of *Cymodocea manatorum* mentioned by most authors (e.g. BøRGESEN, FELDMANN & LAMI, NEWELL et al., VOSS & VOSS) to indicate all seagrasses with narrow leaves (possibly including *Halodule wrightii* and *H. beaudetti*). As a matter of fact *Cymodocea manatorum* does not occur in the West Indies.

24

zonation-pattern for the whole world. The coastal regions studied by them are situated in different parts of the world: Great Britain, South Africa, Indian Ocean and Red Sea, Mauritius, Great Barrier Reef, and North America on its Atlantic and Pacific sides.

The authors came to the conclusion that there are certain widely distributed features in the zonation of intertidal organisms, which may prove to be of general importance. Under the title "The universal features of zonation between tide-marks on rocky coasts" they give their general conclusions, together with a suitable terminology.

They described for *rocky coasts* (Fig. 7):

1. S u p r a l i t t o r a l z o n e; the maritime belt near the sea, above tide-mark, but subject to some marine influence (e.g. spray in rough wheather).
2. S u p r a l i t t o r a l f r i n g e; from the upper limit of barnacles (in quantity) to the nearest higher convenient landmark (e.g. the upper limit of Littorinae or the lower limit of maritime land-lichens or flowering plants). Springtide invades at least the lower part of this zone.
3. M i d l i t t o r a l z o n e; from the upper limit of barnacles (in quantity) down to the upper limit of the next lower zone. This belt tends to be covered and uncovered every day, at least in part.
4. I n f r a l i t t o r a l f r i n g e; from the upper limit of any convenient dominant organism (e.g. *Laminaria, Pyura*) to extreme low-water level at spring tide, or to the lowest level ever visible between waves. This zone is dry only at very low tide in calm weather.
5. I n f r a l i t t o r a l z o n e; from extreme low spring tide to a depth which has yet to be determined — possibly to the edge of the continental shelf or to the lower limit of seaweed vegetation.

The general zonation scheme of the STEPHENSON's has been applied to nearly every part of the world (CHAPMAN & TREVARTHEN, 1953; GUILER, 1953; LEWIS, 1955; FELDMANN, 1955; LAWSON, 1956). However, it has also been criticized, for instance by WOMERSLEY & EDMONDS (1952), and LEWIS (1955). Excellent summaries of the arguments pro and contra are given by SOUTHWARD (1958) and DEN HARTOG (1959).

Particularly discussed by several authors was the terminology. The main objection is against the replacement of the term "sublittoral" by "infralittoral" (WOMERSLEY & EDMONDS, 1952). The resemblance between the words "supralittoral" and "sublittoral" is not a sufficient reason, however, for replacing one of the two terms by another.

DEN HARTOG (1959) criticized the terminology of the STEPHENSON's, because their terms are based on typically English expressions and therefore cannot always readily be translated into other languages. He proposed the term eulittoral instead of midlittoral, and to replace the term "fringe" by a word of Latin origin, viz. "margin". He also rejected the term "zone" in the special meaning of the STEPHENSON's and preferred to speak of "regio".

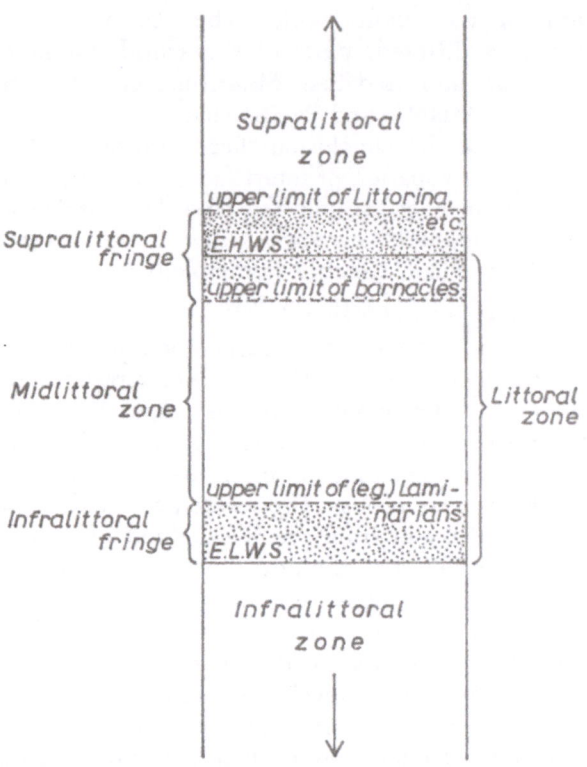

Fig. 7. General zonation scheme for rocky coasts (from T. A. &
A. Stephenson, 1949). E.H.W.S. = extreme high water at spring
tide; E.L.W.S. = extreme low water at spring tide.

It appears that the general zonation scheme by T. A. & A. Stephenson
shows no fundamental differences to the scheme commonly used in
Europe (den Hartog, 1959):

European classification:	Classification of the Stephenson's:
upper supralittoral region	supralittoral zone
lower supralittoral region	supralittoral fringe
eulittoral region	midlittoral zone
sublittoral region	infralittoral fringe
	infralittoral zone

The distinction of a "sublittoral margin" as a separate region seems to be superfluous as, according to DEN HARTOG (1959), it appears to be clear that it is only the uppermost part of the sublittoral region. Under special conditions the vegetation of the sublittoral region may occur somewhat higher than extreme low water at spring tide (especially as a result of wave action). In many cases the vegetation of the sublittoral margin only consists of species of an association which grows to a considerable depth; in a few cases, however, a well-marked association has developed, which clearly deviates from the vegetation at a lower level.

DAHL (1953) has given a general zonation scheme for sandy beaches comparable to that for rocky coasts by the STEPHENSON's (1949). The places studied by him are situated in: Norway, Sweden, Venezuela, Central and Southern Chile.

DAHL's system (1953) only applies to sandy coasts with little organic material, a low percentage of small particles and having a poor fauna. Apparently on sandy beaches a number of zones can be distinguished, comparable with the STEPHENSON zonation scheme.

The following zones are distinguished:
1. S u b t e r r e s t r i a l f r i n g e (Talitrid-Ocypodid belt). — In arctic and antarctic areas this zone is devoid of marine fauna, in temperate regions it harbours mainly Talitrid Amphipods, in warm areas Ocypodid crabs.
2. M i d l i t t o r a l z o n e (*Cirolana* belt). — In warm and temperate areas with Cirolaninae (Isopods), but on the Swedish and British coasts only with a number of Amphipods.
3. S u b l i t t o r a l f r i n g e with a varied fauna. — The Crustaceae no longer dominate to the same extent as in the two higher zones but they still play an important role.

DAHL distinguishes his zones only on the presence of various animals. Macroscopic seaweeds are absent from the littoral region on beaches of clean sand.

In general lines the DAHL's zonation scheme has been confirmed by WOMERSLEY & EDMONDS (1958) with regard to beaches on open shores of Southern Australia, especially for the supralitoral and littoral region. However, DAHL's scheme has also been criticized. GAULD & BUCHANAN (1956) who studied the sandy beaches of the Gold Coast, deny its general value. They distinguish two extra zones, indicated as the *Donax* and *Nerina* zones.

C. ZONATION IN THE CARIBBEAN

The STEPHENSON zonation scheme has been used in the Caribbean by several investigators. In a few cases not only a description of the intertidal zonation on rocky coasts was given, but also the zonation on sandy and muddy shores and of the sublittoral region were studied.

Rocky coasts

T. A. & A. Stephenson (1950) published a survey of the flora and fauna of the tidal zone on rocky shores in the Florida Keys, the coast of which is formed by Pleistocene coral limestone. The zonation is not immediately obvious, but on closer examination it can be correlated with the zonation in other parts of the world.

On the Keys the rocky parts of the tidal zone show the following structure, though the details vary from Key to Key, and local modifications exist in relation to topography, depth, sedimentation, and wave action. "The most conspicuous intertidal feature is a low but well-marked platform of rock (upper platform) extending from the edge of the land vegetation towards the sea. This varies in width and in the sharpness of its seaward edge. At the foot of it, to seaward, lies a much lower, less strongly marked lower platform, variable in development, often much interrupted, and sometimes absent. More seaward again is the reef flat, a low-lying area in which rocky patches alternate with sand, mud and gravel; this is commonly covered by shallow sea at low water, and its higher parts emerge as banks at the lowest tides". The slope from the upper to the lower platform may be abrupt, in several cases even a niche is formed, in other cases a vertical wall. This description of the coastal forms of the Florida Keys shows much resemblance to that of de Buisonjé & Zonneveld (1960) concerning the limestone cliffs of Curaçao, Aruba and Bonaire.

The supralittoral margin (in connection with the wide, almost flat upper platforms of many of the Keys) is well developed and may be subdivided into subzones, mainly based on a varied population of snails. The zones of the supralittoral margin are indicated as white, grey and black zones respectively.

The white zone is rarely and irregularly wetted by the waves and even then only in the lower parts. In many places the gastropod Tectarius muricatus is found. The degree of development of the land vegetation appears to be variable.

The grey zone lies seaward to the white zone and may be wetted occasionally by the waves at spring tide. The vegetation of this zone may include three species of mangroves (Laguncularia racemosa, Avicennia germinans and Rhizophora mangle). Also Sesuvium portulacastrum, Batis maritima, Salicornia perennis may be present. This is the first zone with macroscopic marine algae, especially Bostrychia spp., forming moss-like growths on the roots of mangroves and in rock cavities. Characteristic snails are Littorina ziczac, Tectarius muricatus, T. tuberculatus, Echininus nodulosus, Nerita versicolor and Nerita peloronta (the last one reaching its maximum in the grey zone).

The black zone is regularly and completely wetted by the waves at spring tide. The surface of the rock is uneven and dissected. The dark colour is caused a.o. by the bluegreen algae Entophysalis deusta (= granulosa) and Brachytrichia quoyi together with the green alga Tellamia intricata. Littorina ziczac, Tectarius tuberculatus and Nerita versicolor reach their maximum in the black zone. Nerita tessellata, although characteristic for the yellow zone, may also be found. Nerita peloronta, most abundant in the grey zone, may still be present. The only common macroscopic marine algae are Bostrychia binderi and B. tenella, forming a moss-like growth, frequently accompanied by other small forms, e.g. Polysiphonia howei and Murrayella periclados in rock cavities (in the seaward part of the black zone).

The e u l i t t o r a l   r e g i o n includes the rather abrupt transition from the upper to the lower platform and is, in relation to the small tidal differences, rather narrow. Barnacles are only locally present. This region is called *yellow zone.*

Although there are many overhanging rocks in this zone, no special cryptofauna or flora have developed; mostly a moss-like growth of *Bostrychia* is found, and small mussels may be numerous. Other algae in this zone are *Anadyomene stellata, Cladophoropsis membranacea, Catenella repens, Centroceras clavulatum, Ceramium fastigiatum* and *Polysiphonia howei.* Together they form a moss-like algal growth. It is not clear what causes the yellow colour of this zone.

In many places the yellow zone can be sub-divided into an upper yellow zone and a lower yellow zone. The u p p e r   y e l l o w   z o n e is characterized by the barnacle *Chthamalus* and a moss-like growth of *Bostrychia,* the lower yellow zone by the green alga *Valonia ocellata* and the tube-mollusc *Spiroglyphus irregularis.* The yellow zone also contains many molluscs. *Nerita tesellata* and *Nerita versicolor* are abundant. Furthermore the mussel *Mytilus exustus* and a large chiton, *Acanthopleura granulata* are characteristic. In the l o w e r part of the y e l l o w   z o n e the algal growth in places may be conspicuously developed, forming yellowish-brown beard-like masses of *Centroceras clavulatum, Polysiphonia ferulacea, Polysiphonia sphaerocarpa* (prox.) and *Spyridia filamentosa.*

The lower platform is located below the yellow zone. It is not easy to define exactly where the lower platform ends and the reef flat begins. The lower platform is fully exposed to the air at mean low water at spring tide; on the other hand it is regularly washed by the waves and it may be distinguished as a definite and separate region (sublittoral margin).

The most distinctive feature is the occurrence of a low yellowish carpet of *Laurencia papillosa,* mostly growing together with a number of other algal species. *Valonia ocellata* is abundant both on the lower platform and in the lower yellow zone. The same applies to *Spiroglyphus irregularis*; this species however, has a sharp lower limit. In several places on the lower platform large cushion-like masses of *Halimeda opuntia* are found, locally with large patches of *Zoanthus sociatus.* Many other organisms may be present, among them the short-spined urchin *Echinometra lucunter,* in cavities of the rock and under pebbles.

Finally T. A. & A. STEPHENSON (1950) give detailed lists of organisms present on the reef-flat.

Several parts of the reef-flat emerge from the sea at extreme low tides and therefore belong to the sublittoral margin. Most parts, however, are permanently submerged and belong to the sublittoral region proper.

Many of the organisms of the reef-flat occur in large numbers or form definite patches. This is true for the flowering plants, such as *Thalassia* but also for mussels (*Mytilus exustus*), corals (*Porites*) and several species of algae (*Jania capillacea, Spyridia filamentosa, Halimeda opuntia* and branched Corallinaceae).

Along the oceanic side of the Florida Keys a submarine bank is found, the bottom of which is formed in part by sand and mud and then covered by a dense *Thalassia* growth. Other areas have a rocky bottom supporting corals and gorgonids.

Much attention has been paid to the publication of T. A. & A. STEPHENSON because it is the first description for the tropical western part of the Atlantic Ocean, giving a survey of the pattern of plant and animal zonation from a general biological point of view.

Their investigations have been tested and supplemented for other parts of the Caribbean. Next to the eulittoral region, the sublittoral region also was studied (VOSS & VOSS, 1955; RODRIGUEZ, 1959; NEWELL, IMBRIE, PURDY & THURBER, 1959; LEWIS, 1960).

VOSS & VOSS (1955) published an ecological survey of SOLDIER KEY, Biscayne Bay, Florida. In the shallow coastal waters the following zones are distinguished: *Echinometra* zone, *Porites*-coralline zone, *Thalassia* zone, and *Alcyonaria* zone.

Echinometra zone. — At the boundary between upper and lower platform the rocks are inhabited by the urchin *Echinometra lucunter* to a depth of about 45 cm. The starfish *Echinaster sentus* is abundant. Between and underneath rocks and boulders only few organisms are present in this zone.

Porites - coralline zone. — The eastern coast of Soldier Key shows a dense growth of non-attached plants of *Jania*, *Amphiroa* and *Goniolithon*, together with the coral *Porites furcata*, and extensive patches of *Thalassia* and *Syringodium*. Several green algae, e.g. *Penicillus* and *Halimeda*, may be also found. This zone is about 15 m wide; the water is 30 to 60 cm deep.

Thalassia zone. — At the lower level a *Thalassia* zone is well developed on the sandy bottom and is about 30 to 45 m wide. Animals in this zone are sponges, molluscs and corals, the sea cucumber *Holothuria floridana*, the urchins *Lytechinus variegatus*, and *Tripneustes esculentus*.

Alcyonaria zone. — Downward of the *Thalassia* zone a rich growth of gorgonians is found, particularly on rocky bottoms. It is one of the most remarkable features of the West Indian area. Soldier Key is one of the northernmost places in which it has been observed.

According to the Voss'es the four zones may be considered characteristic also for the other islands of the Florida Keys.

The investigations of RODRIGUEZ (1959) were undertaken in another part of the Caribbean, i.e. the island of MARGARITA off the north coast of Venezuela . He described the marine communities on rocky, sandy and muddy shores. On rocky shores the following zones were distinguished: *Littorina* zone, Balanoid zone, *Echinometra* zone, *Thalassia* zone, and *Alcyonaria-coral* zone.

In general lines the same communities are found on Margarita as described for the Florida Keys, with the exception of a separate *Porites*-zone. The *Alcyonaria* zone on rocky bottom appears to be "equivalent to the formations of *Thalassia* and mangroves of the sandy bottoms".

On the rocky coasts of Margarita a well-developed algal belt is mostly present in the upper part of the sublittoral region. Development and composition have a distinct relation to wave action. In exposed places *Sargassum filipendula* and *Pterocladia* spec. are dominant, in more protected places they are replaced by *Ulva fasciata* and *Laurencia papillosa*. In places with strong tidal currents the *Ulva — Laurencia* community is replaced by a *Grateloupia cuneifolia — Laurencia papillosa* community.

In the sublittoral region the rocks are generally covered by crust-like Lithothamnia. Although the scheme given by RODRIGUEZ (1959) in several cases shows marked differences to his own descriptions for certain observation stations (compare for instance fig. 15 with the column St. 2 SB 1 in fig. 20), it may be concluded that his observations agree to a great extent with those of the STEPHENSONS.

NEWELL, IMBRIE, PURDY & THURBER (1959) described the communities and bottom facies of GREAT BAHAMA BANK. The rocky coast in most places of the Bahamas shows a marked uniformity, both in physical and in ecological respects. Four distinctive communities were recognized on the coral limestone substrate: rocky shores, infratidal rocky prominences, coral reefs, and rock pavements of submerged marine terraces. Following the STEPHENSONS (1950), they described on rocky shores, in descending order a w h i t e, a g r e y, a b l a c k and a y e l l o w z o n e, the last one subdivided into an u p p e r and a l o w e r y e l l o w z o n e.

Remarkable species for the upper yellow zone are *Bostrychia, Chthamalus* and *Nerita versicolor*; the lower yellow zone is characterized by the vermetid gastropod *Spiroglyphus irregularis* and a boring barnacle, *Lithotrya dorsalis*. The list of organisms given for the different zones appears to be in close accordance with the observations of the STEPHENSONS (1950).

Immediately below the lower yellow zone NEWELL et al. distinguished a "c o r a l- l i n e l i p z o n e". Just above the level of low spring tides there is a dense growth of the encrusting corallinaceae *Porolithon pachydermum*. Other organisms in this zone are *Chiton viridis, Spiroglyphus irregularis, Thais deltoides, Livona pica* and *Echinometra lucunter*. The zone therefore shows great similarity to the *Echinometra* zone as described by Voss & Voss (1955) and RODRIGUEZ (1959).

On i n f r a t i d a l r o c k y p r o m i n e n c e s in turbulent waters in the sub- littoral region NEWELL et al. distinguished the "Millepora community". The most distinctive and common species are *Valonia ocellata, Cladophoropsis membranacea, Turbinaria turbinata, Padina sanctae-crucis, Laurencia papillosa* and *Gonolithon solu- bile*. Also numerous animal organisms are found, such as *Zoanthus sociatus, Millepora, Gorgonia flabellum, Echinometra lucunter, Diadema antillarum, Livona pica, Thais rustica, Lithophaga bisulcata* and *Panilurus argus*. The STEPHENSONS (1950) considered many of these species characteristic of the lower platform. However, according to NEWELL et al. (p. 211), there are also a number of equally important differences, for instance the presence of *Porolithon, Turbinaria, Padina, Millepora, Siderastrea* and *Gorgonia*.

C o r a l r e e f s are generally found on shoals along the margins of submerged erosion platforms, particularly on the windward side of the islands. The authoıs distinguished an *Acropora palmata* community, with abundant *Montastrea annularis, M. cavernosa, Siderastrea siderea, Diploria labyrinthiformis, Acropora cervicornis* and *Porites porites*. When the water becomes very shallow *Millepora alcicornis* dominates.

A distinctive feature of the Great Bahama Bank is the presence of extensive level r o c k y b o t t o m s, swept clear of sediments by waves and currents. These sub- merged erosional terraces support only a sparse growth of algae, sponges and a few massive coral species (*Montastrea* and *Diploria*). At a greater depth, below the influence of storm waves, the most characteristic animals are representatives of the family Plexauridae (seawhips and sea-fans). In many places an ephemeral thin blanket of sand covers the rock surface, bearing the "Plexaurid community". Once established, it can withstand the constant shifting of the bottom sediments. However, when the sediment cover exceeds about 20 cm, the Plexaurid community is replaced by other organisms, such as calcareous green algae of the order Siphonales and also *Thalassia* and *Syringodium*.

The zonation of the intertidal fauna of rocky shores of BARBADOS was described by LEWIS (1960). In spite of the fact that LEWIS was familiar with the STEPHENSONS publications (1949, 1950), he gave his description another form.

In the cliff of coral limestone, in typical cases a small undercut has been formed somewhat below mean low water. Also at the mean high water mark a niche is present, usually somewhat larger (Fig. 8). The cliff face of the overhanging rock is practically vertical when the coast is steep. When the coast is low a sharp ridge is formed. The upper side of the rocks is nearly horizontal and has a very rough surface.

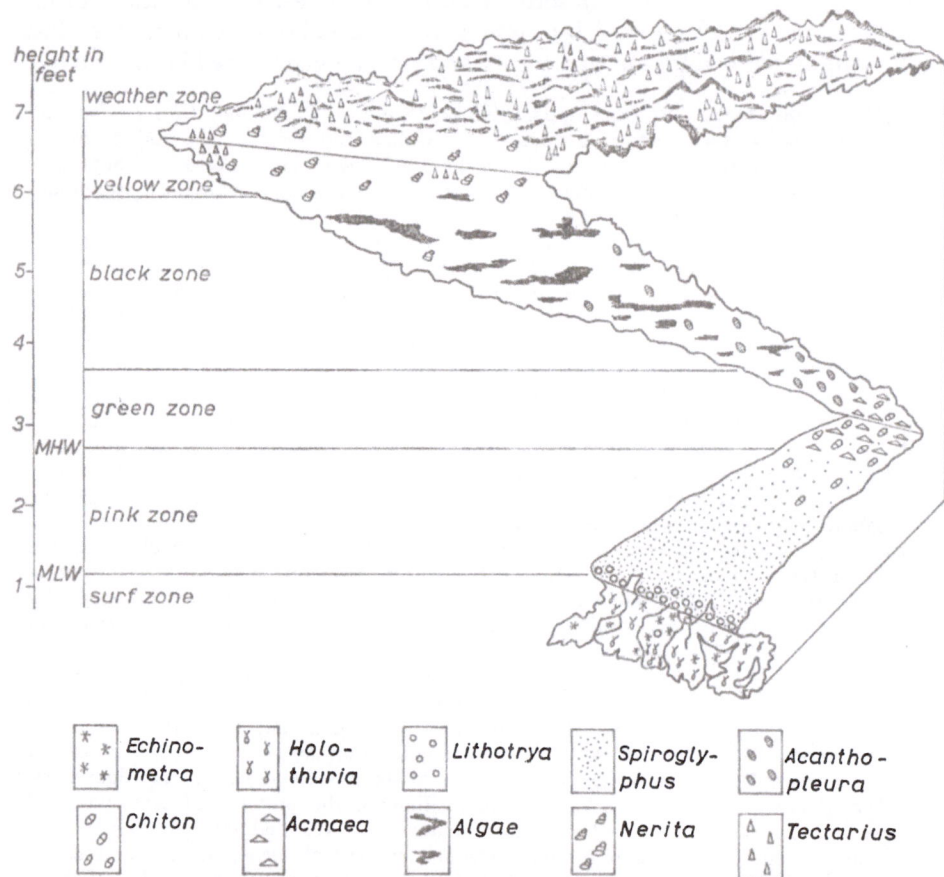

Fig. 8.   Diagrammatic profile of a typical cliff coast (after Lewis, 1960).

In broad outline the coasts of Barbados may be compared with those of Curaçao, Aruba and Bonaire, as described by DE BUISONJÉ & ZONNE-VELD (1960).

Several zones were recognized by LEWIS (1960), and clearly defined by colour changes on the rock. He distinguished 6 different zones, viz. weather zone, yellow zone, black zone, green zone, pink zone and surf zone. With exception of the surf zone they are all situated in the intertidal region.

Weather zone. — This is the uppermost zone of the intertidal region. On low-cliffed coasts, land vegetation reaches into this zone (*Ipomoea pes-caprae*). The colour of the rock is dark, due to microscopic algae. Common molluscs are *Tectarius tuberculatus, Littorina ziczac, Nerita peloronta* and *Nerita versicolor*.

Yellow zone. — This zone corresponds approximately to the outer ridge of the major undercut. The surface of the rock is rough. The yellow discoloration of the surface is due to the same microscopic organisms which are found in the weather zone; here, however, they are not dry but wet. Common molluscs are *Tectarius tuberculatus, Littorina ziczac, Nerita peloronta* and *Nerita versicolor*.

Black zone. — The black zone is found under the upper edge of the major undercut of the cliff. The surface of the rock in this zone is rough and pitted. Common are *Bostrychia tenella* and *Polysiphonia howei*, the last one somewhat below *Bostrychia*. Sometimes these algae form a dense vegetation excluding all animals; usually, however, small patches of algae are formed. The common animals are *Acanthopleura granulata* and *Thais patula*.

Green zone. — The colour of the rock is a light yellow or light green, caused by microscopic boring algae. The surface of the rock usually appears smooth. The mean high water mark falls within the limits of this zone. Common organisms are *Acanthopleura granulata, Thais patula, Thais floridana* and *Acmaea jamaicensis*.

Pink zone. — The pink colour of this zone is one of the most conspicuous features of the rocky shores of Barbados. It is due to the growth of lime-encrusting coralline algae (Lithothamnia). The upper limit lies just below the high water mark. The surface of the rock usually is smooth. The most characteristic animal is *Spyroglyphus irregularis*. Also a number of molluscs is present, e.g. *Fissurella barbadensis, Leucozonia ocellata, Thais floridana, Chiton marmoratus* and actinians as *Bunodactis* and *Bunodosoma*.

Surf zone. — The surf zone forms a relatively narrow band between mean low water and approximately mean low water springs. It is rarely completely dry; even when there is little wind the rocks are constantly wetted. In some places the surface of the rock has deep fissures and cracks; in other places it is rather smooth. The surf zone has a dense growth of algae. The common organisms are *Echinometra, Fissurella barbadensis, Holothuria glaberrima, Lithotrya dorsalis, Spirobranchus giganteus, Bunodosoma cavernata*, sponges, Bryozoans, *Sargassum* spp. and other algae.

The zones distinguished by LEWIS (1960), on account of their dominating organisms, can be easily compared with the zones described by T. A. & A. STEPHENSON (1950). The weather zone and the yellow zone of LEWIS correspond to their white, grey and black zone of the upper platform (supralittoral margin). The black zone and the green zone correspond to their upper yellow zone, the pink zone to the lower yellow zone. The surf zone finally corresponds to the lower platform (sublittoral margin). The colour of the rocks cannot be used for correlating the different zones since it varies considerably from one place to another.

The publications mentioned before all deal with rocky coral limestone shores. Only RODRIGUEZ did not mention the nature of the rock clearly. From this survey it appears that the general zonation scheme of the STEPHENSONS (1949) is valid for different parts of the Caribbean, even when they are situated quite a distance apart. It is also shown that most of the organisms, characteristic for a definite zone, occur all over the Caribbean area. This means a considerable simplification in comparing publications and observations.

The STEPHENSONS (1950) also dealt with the zonation on a vertical wall. In general there appears to be a distinct resemblance to naturally rocky coasts. There are also some differences, particularly in the zonation of *Nerita* species.

LEWIS (1960) described the zonation on beachrock for one place on Barbados. He distinguished: surf zone, *Fissurella* zone, boulder zone, and *Neritina-Ulva* zone.

The surf zone roughly corresponds to the same zone on cliff coasts. It has deep fissures, running at right angles to the coast. *Echinometra* is abundantly present. The layer of coralline algae is only thin and irregularly developed.

The *Fissurella* zone is an almost horizontal platform, with several cracks running parallel to the shore. The algae form a very dense, short and moss-like vegetation. *Fissurella barbadensis* appears to be numerous.

The boulder zone is found behind the lightly sloping platform of beachrock. Even at low tides this zone is covered with water that is regularly renewed through cracks in the rock. Common are juvenile *Tripneustes esculentus* and *Diadema antillarum*.

In the *Neritina-Ulva* zone both *Neritina pupa* and *Ulva lactuca* are abundant. *Littorina meleagris* and *Nerita tessellata* may also be found.

These observations of LEWIS (1960) for beachrock in several respects show a resemblance to the scheme of zonation for coral limestone.

The surf zone of both coastal forms is comparable, the *Fissurella* zone on beachrock shows affinities to the pink zone on coral limestone.

The boulder zone, however, cannot be easily compared; the tops of the highest boulders of this zone may reach to a height of 30 cm above the level of mean low water. This means that a number of the boulders become completely dry. They are partly covered by a dense moss-like algal vegetation. On the sides of the rocks numerous specimens of *Fissurella barbadensis*, *Acmaea jamaicensis* and *Nerita tessellata* occur; on the lower part, dry at low water, a dense population of *Spirobranchus giganteus* is observed. The boulder zone therefore represents a number of different zones.

The *Neritina-Ulva* zone finally passes at its landside into a sandy beach.

The coasts of the islands of ST. MARTIN, ST. EUSTATIUS and SABA for the greater part consist of volcanic rock, often with large boulders, and several habitat factors may vary greatly. The degree of exposure to wave action is different on the front and on the back of the boulders; there are also variations in exposure to sunshine and to the erosive action of water loaded with sand and gravel. According to the nature of the rock the seaweeds may find varying possibilities for attachment to the rock.

The publications dealt with in the foregoing pages hardly pay attention to these factors. Only LEWIS (1960) discussed a "Comparison of stations and influence of physical factors".

SANDY AND MUDDY BEACHES

The Caribbean Sea has a great number of beautiful beaches. Many of them are situated on open, exposed coasts, others are found at the leeward side of the islands and in lagoons.

For the Danish West Indies (Virgin Islands) Børgesen (1909) distinguished the following coastal vegetations:

A. The hydrophyte vegetation
    I. The muddy and sandy soil vegetation
        1. The seagrass (and algae) formation

B. The halophyte vegetation
    I. The muddy soil vegetation
        1. the mangrove formation
        2. the *Salicornia* formation
        3. the *Conocarpus* formation
    II. The sand strand vegetation
        1. the *Pes caprae* formation
        2. the *Tournefortia* formation
        3.   the *Coccoloba* — machineel formation
    III. The rocky coast vegetation

The s e a g r a s s (and algae) f o r m a t i o n includes several species of phanerogams: *Syringodium filiforme, Thalassia testudinum, Halophila baillonis* and *H. aschersoni*. Among these seagrasses many algae of the genera *Caulerpa, Penicillus, Udotea* and *Halimeda* are present. These algae are fixed in the sandy or muddy bottom by means of small rhizoids. The lime-encrustating algae produce a great deal of the material forming the sea bottom; in some cases the bottom nearly completely consists of grit of *Halimeda* remnants. The seagrasses go to a depth of 5—6 fathoms, the seaweeds much deeper.

The m a n g r o v e f o r m a t i o n was described by Børgesen (1909) as "a formation of tree-like evergreen plants, growing on the sheltered shores, partly in shallow, salt or brackish water, partly on low-lying soil which is comparatively rarely, sometimes perhaps never covered by salt or brackish water". Three species of mangroves occur in this formation: *Rhizophora mangle, Avicennia germinans* and *Laguncularia racemosa*. Børgesen (1909) is of the opinion that the mangroves begin to grow on a rocky or sandy bottom. Afterwards mud is deposited between the roots of the mangroves. When the trees grow near the mouth of the lagoon, this lagoon can be completely shut off from the sea by deposition of mud.

The S a l i c o r n i a f o r m a t i o n is found on marshy mud flats where often *Laguncularia* is present. The vegetation is in most cases very sparse and the clay bottom is visible between the plants. Common species of the *Salicornia* formation are *Salicornia ambigua, Batis maritima* and *Sesuvium portulacastrum*.

Landward to the mangrove formation, at a higher level and with less salt in the bottom, the C o n o c a r p u s f o r m a t i o n is found. It is only submerged at extremely high water and is even then not always so.

The bottom is clayish, sometimes with a little sand. The most common species of this vegetation are *Conocarpus erecta*, *Annona palustris*, *Bucida buceras* and *Acrostichum aureum*.

BøRGESEN's sand strand vegetation, which grows on sandy bottoms at a higher level, will not be discussed here. Likewise the vegetation of rocky coasts will be left out of consideration. However, we will dwell at length on the first two formations distinguished by BøRGESEN (1909), who, moreover, in 1911 gave a more detailed description of these formations in his paper on "The algal vegetations of the lagoons of the Danish West Indies". The lagoons, "areas of shallow water which by coral reefs or most

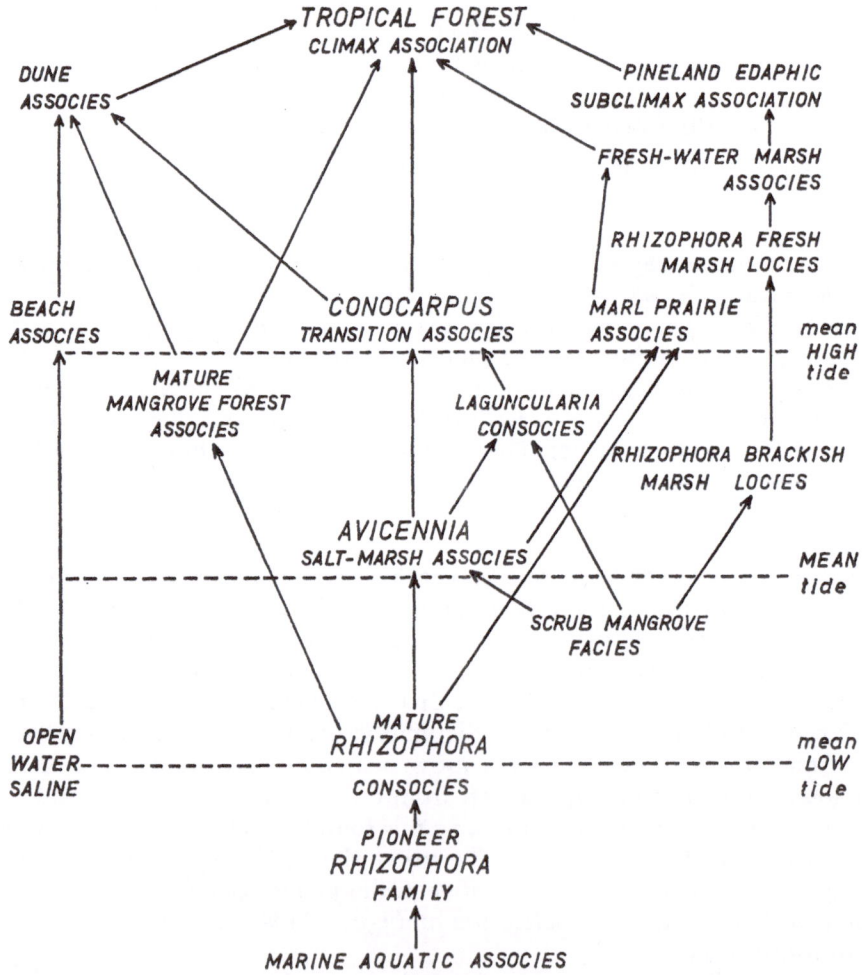

Fig. 9. Successional relations of mangrove communities and some of the associated plant communities. Approximate tide levels are indicated (after Davis, 1940).

often low and narrow tongues of land are more or less separated from the sea", have special conditions of life, which will be described below.

The shore of the lagoons is generally bordered by a dense vegetation of m a n g r o v e s, on which much has been published (e.g. GERLACH, 1958, and DAVIS, 1940). STOFFERS (1956) gave a description of the "mangrove woodlands" in the Netherlands Antilles. Figure 9 is taken from DAVIS (1940) and gives an illustration of the succession of mangrove communities and related vegetations. From the mean tide levels a clear picture is obtained of the mutual relations of the communities of plants and also of the time of submergence for different zones.

In the shallow, quiet water of the lagoon, much mud is gradually deposited and the mangrove vegetation considerably accelerates this process. In northern seas we can hardly find any growth of algae under such conditions. In the tropics, however, the contrary appears to be true. Not only do we find a rich and well-developed algal growth on the roots of mangroves, which in a certain way may be regarded a substitute for a rocky substrate, but the soft sandy and muddy bottom of the lagoons is also covered with a dense vegetation (BØRGESEN, 1909, 1911). An important condition for this growth is that the water must be sufficiently salt and clean, which generally is far from being the case.

When the opening of the lagoon to the sea is narrow and shallow (and mangroves are often a contributing factor), the renewal of water is slow. Then only in the foremost part of the lagoon algal growth is possible.

The richly developed plants and animals on the roots of mangroves, especially of *Rhizophora*, have drawn the attention of many authors. These organisms are found both in the sublittoral region and at a higher level, where, in the shadow of the foliage, their growth is possible. Other authors giving information on these algal vegetations, are e.g. FELDMANN & LAMI (1936), POST (1936), TAYLOR (1954, 1959, 1960), ALMODÓVAR & BIEBL (1962), BIEBL (1962) and CONOVER (1964).

In the supralittoral region, on the trunks of *Rhizophora* a golden-yellow cuff is formed by the lichen *Pyrenula cerina*, which in some places occurs together with grey specimens of *Opegrapha* (FELDMANN & LAMI, 1936). At a lower level, shaded by the foliage on the roots of *Rhizophora* (and to a much lesser degree also on the pneumatophores of *Avicennia*), a dense growth of algae may be found. This algal vegetation has a sharp upper and lower boundary.

The roots of *Rhizophora* in the littoral zone are mostly densely covered with *Bostrychia binderi* and *B. tenella*, together with *Caloglossa leprieurii* and *Catenella opuntia*. This combination of sciaphilous algae, indicated as *Bostrychia-Caloglossa* association (*Bostrychietum*) by POST (1936), is found everywhere in the world under the same conditions. In temperate areas several species disappear and only *Bostrychia* and *Catenella* remain.

The sublittoral region of the mangrove roots is characterized by many other species. *Caulerpa verticillata* may form dense vegetations, often together with a number of other algae. Several authors give an enumeration of the algae found on the roots of mangroves. Abundant are for instance *Murrayella periclados, Acanthophora spicifera, Centroceras clavulatum, Ceramium fastigiatum, Hypnea musciformis, Laurencia obtusa,*

Polysiphonia havanensis, Polysiphonia ferulacea, Wrangelia bicuspidata, Wurdemannia miniata, Acetabularia crenulata, Dichotomosiphon pusillus, Enteromorpha flexuosa, Neomeris annulata, Rhizoclonium hookeri, Valonia spp., *Caulerpa* spp. Representatives of the last-mentioned genus also inhabit the sandy mud between the roots of the mangroves, such as *Caulerpa verticillata, C. sertularioides* and *C. taxifolius*. Many of the algal species mentioned already are also abundant on rocky shores and therefore cannot be considered as typical of mangrove vegetations.

The degree of development of these algae, and also their presence or absence, is closely correlated with local features, especially the renewal of water. The algae may be covered with a thick layer of detritus. This is partly an explanation for the fact that usually few algae may be found on the pneumatophores of *Avicennia: Rhizoclonium hookeri* at high water mark and *Bostrychia* in the littoral region (TAYLOR, 1960).

Few publications deal with the organisms of mangrove vegetations from a general biological point of view. The STEPHENSONS (1950) did not give a separate review of the mangroves, they only mention casually the occurrence of mangroves when they are dealing with the grey zone. GERLACH (1958) did not pay much attention to the algal vegetations on the roots of the mangroves; however, he enumerated the animal organisms typical of different zones, with stress on the Nematodes. Also RODRIGUEZ (1959), NEWELL et al. (1959) dedicated a few lines to the plants and animals in different zones on the roots of mangroves. However, it appears difficult to obtain a general picture from the publications mentioned.

Mangrove vegetations generally show considerable local variations in habitat factors. This is clearly indicated by GERLACH (1955) in his ecological survey of the nematodes in mangrove vegetations along the coast of Brasil.

The formation of s e a g r a s s e s (and algae) distinguished by BØR-GESEN (1909) must also be reviewed. In several publications (1900, 1909 and 1911) he described this vegetation in the Danish West Indies (Virgin Islands). Other authors (FELDMANN & LAMI, 1937; TAYLOR, 1954, 1960) added information on other parts of the Caribbean area. However, all these authors focused their attention on seagrasses and algae and did not pay much attention to other organisms.

A more general description was given by NEWELL, IMBRIE, PURDY & THURBER (1959), in a review of the "sediment bottom habitats and communities" of Great Bahama Bank.

These authors distinguished the following communities:
1. *B e a c h* c o m m u n i t y.
2. *S t r o m b u s   s a m b a* c o m m u n i t y. — This community is found on unstable sand bottoms, generally deeper than 2 fathoms, and it is characteristic of relatively clear turbulent waters. — Organisms of this community are *Udotea flabellum, Halimeda* spp., *Acetabularia* spec., *Padina sanctae-crucis* and *Laurencia intricata* and furthermore *Syringodium* and *Thalassia* (forming sparse patches), the foraminifer *Rotalia rosea*, polychaete worms, the echinoid *Mellita sexiesperforata*, many species of Pelecypods, and the gastropod *Strombus samba*.
3. *T i v e l a* c o m m u n i t y. — This is the community of unstable oolite sand bottoms which are devoid of plant and animal life as a result of the mobility of the

substrate. — *Thalassia* and *Syringodium* form a sparse growth. The most abundant forms are *Syringodium*, the pelecypod *Tivela abaconis* and the echinoderm *Oreaster reticulatus*.

4. *S t r o m b u s   c o s t a t u s* community. — This is a rich and diverse community of the shelf lagoon, with a stable sand bottom generally less than 1.5 fathoms deep. — The community is particularly characterized by many echinoderms and molluscs living in a moderately heavy plant cover of seagrass and algae. *Goniolithon strictum*, a coralline alga, is abundant on shallow grass bottoms in protected places; locally it may be an important sediment former. Between the beds of seagrasses, mainly formed by *Thalassia*, several species of algae are found, e.g. *Halimeda*, *Penicillus pyriformis*, *Acetabularia crenulata*, *Dictyosphaeria cavernosa*, *Rhipocephalus phoenix*, *Laurencia intricata*, *Goniolithon strictum*. Abundant animals are the echinoderms *Clypeaster roseus* and *Oreaster reticulatus*, the corals *Porites furcatus* and *Manicina areolata*, and the gastropod *Strombus costatus*.

5. *D i d e m n u m* community. — This is a community of a stable bottom of muddy sand, with hypersaline water. — The most conspicuous surface-dwelling organisms are *Caulerpa paspaloides*, the yellow-coloured sponge *Verongia fistularis*, and the greyish-white colonial tunicate *Didemnum candidum*.

6. *C e r i t h i d e a* community. — This community is considerably less diverse than the *Didemnum* community. — *Batophora ocrstedi*, a green alga, is abundant in this poikilohaline habitat. The molluscs *Cerithidea costata* and *Pseudocyrena colorata* seem to be limited largely to this habitat.

7. *M a n g r o v e   c o m m u n i t y*. — The mangrove community is formed on intertidal mud and muddy sand in water not deeper than about 3 feet at high tide. — The trees play an important role in accumulating the bottom particles. The bottom is rich in H_2S. On the roots of the mangroves various filamentous algae are found. *Batophora* is common and gastropods live among the filamentous algae. — The tidal creeks between the mangroves mostly have a muddy bottom with their own characteristic species.

The communities, as described by NEWELL et al. (1959) appear to be closely related to the bottom facies. The plants on these sediment bottoms are well anchored by means of creeping stems *(Thalassia, Syringodium)* or bundles of rhizoids (green algae of the order Siphonales). Between the plants many animals are found, partly buried in the bottom (molluscs, worms, crustaceans).

This detailed account of the communities of the Great Bahama Bank gives more information, especially of the *Thalassia* zone, than the description by Voss & Voss (1955) and RODRIGUEZ (1959) of sandy bottoms in the sublittoral region. *Thalassia* and also *Syringodium* may grow under quite different circumstances. They are found in nearly all the communities distinguished by NEWELL et al. (1959), their degree of development, however, is variable.

In 1959 RODRIGUEZ also gave a further study on the *Thalassia* formation, starting, however, from other basic ideas than NEWELL et al. (1959).

On sandy coasts of Margarita he distinguished the following communities:

   I. *Ocypode* zone
  II. *Talitridae* zone
     1. Excavating forms
     2. Forms living upon debris
 III. *Donax-Tivela* zone

IV. *Emerita* zone
V. *Mellita* zone
VI. *Thalassia* formation
    1. Forms living on the tips of the leaves: *Colpomenia-Ulva* association.
    2. Forms buried within the leaves and roots: *Thyone* — pelecypod association.
    3. Forms living on patches of *Thalassia*: *Strombus-Oreaster* association.

These observations partly correspond to those of DAHL (1959). The *Thalassia* formation is seen by RODRIGUEZ (1959) "as a biocenosis which is a unit in itself and thus can be considered, together with the mangroves as one of the two formations of sandy bottoms. The bathymetrical limits of *Thalassia* range from L.W.S. to a few meters deep. On the upper limit of their distribution, the limiting factor is apparently the degree of exposure to wave action".

The algae which may be found as epiphytes on *Thalassia* are summarized in a recent paper by HUMM (1964). CONOVER (1964) published some information on this subject also.

RODRIGUEZ (1959) recognized in a single locality with a muddy bottom the following communities: 1. *Uca-Tellina* association; 2. *Syringodium* consocies; 3. *Squilla* — polychaete (?) association. He concludes: "The composition of sandy shores varies greatly both in number of species and of individuals, according to factors that were not apparent". — Muddy shores have also a fauna of pelecypods and crustaceans, but the species are quite different from those in the sand. There is a predominance of crabs (*Uca*) and the horizontal zones are not so well marked. — "In general the communities of tropical muddy shores are very poorly known".

CHAPTER IV

SURVEY OF THE OBSERVATIONS ON THE ALGAL VEGETATIONS
OF ST. MARTIN, ST. EUSTATIUS AND SABA

A. GENERAL REMARKS

In the same way as was done with the survey of the literature on zonation in the Caribbean, the algal vegetations of rocky shores and of sandy and muddy shores will be also discussed separately.

The localities on St. Martin, St. Eustatius and Saba are numbered in the sequence of stays. On St. Martin 35 samples were taken, on St. Eustatius 6, and on Saba 8 (see Figs. 10, 4 and 3 respectively). These localities will be indicated by the name of the island (abbreviated to St. M. or St. Eust.), followed by the relevant number, for example St. M. 17 for Guana Bay on St. Martin.

In the maps the station numbers of WAGENAAR HUMMELINCK are also included; they consist of 3 of 4 figures.

Names of well-known species are not always repeated in full, if errors are excluded, e.g. *Rhizophora* (= *R. mangle*), *Avicennia* (= *A. germinans*, syn. *A. nitida*), *Laguncularia* (= *L. racemosa*); *Thalassia* (= *T. testudinum*), *Syringodium* (= *S. filiforme*), *Turbinaria* (= *T. turbinata*), *Batophora* (= *B. oerstedi*), *Pocockiella* (= *P. variegata*), *Tectarius* (= *T. tuberculatus*, syn. *Nodilittorina tuberculata*), *Littorina* (= *L. ziczac*), and *Echinometra* (= *E. lucunter*).

The word Chiton is used in a general sense (the most common species is *Acanthopleura granulata*); Lithothamnia indicates all crustlike, pink-coloured non-articulate forms of Corallinaceae.

B. ALGAL VEGETATIONS OF ROCKY SHORES

In the algal vegetations on the rocky coasts of St. Martin the following types can be distinguished:
1. algal vegetations on coral limestone.
2. algal vegetations on rocks of the Low Lands formation.
3. algal vegetations on rocks of the Point Blanche formation.
4. algal vegetations on dolerite.
5. algal vegetations on diorite.
6. algal vegetations on beachrock.
As to the coast of the islands of St. Eustatius and Saba, which consists mainly of material of volcanic origin, we have in addition:
7. algal vegetations on andesite.

Firstly the heavily exposed places will be discussed, then the moderately exposed coasts and finally the sheltered sampling spots.

As the coast is generally formed by boulders of various sizes, the degree of exposure is difficult to determine. Great differences can be observed

41

between the front side and the back side of the same boulder. The insolation and the scouring effect of sand and gravel may also have different effects. The relation between exposure and zonation is most striking on steep coasts which consist of massive rocks.

B. 1. ALGAL VEGETATIONS ON CORAL LIMESTONE (*St. Martin*)

On St. Martin coral limestone is found in several places along the shore and under varying circumstances.

Legenda:
1. Guana Bay, 13.V.1958.
Heavy swell; littoral region and upper part of sublittoral region. — Collecting number: St. Martin 17 (Pls. IVa and VIIIb; Fig. 11).
2. Guana Bay, 13.V.1958.
Free floating algae in front of sandy beach, south of coast of coral limestone. — Collecting number: St. Martin 19.
3. Point Blanche Bay, 14.V.1958.
Rather heavy swell; littoral region and upper part of sublittoral region. — Collecting number: St. Martin 20 (Pl. IVb; Figs. 12 and 13).
4. Little Bay, eastern part, 30.IV.1958.
Little swell; littoral and sublittoral region. — Collecting number: St. Martin 5 (Pl. IIIb; Fig. 14).
5. Little Bay, eastern part, 30.IV.1958.
Rocks of coral limestone at sea-level, constantly washed by water with sand; littoral region and upper part of sublittoral region. — Collecting number: St. Martin 4 (Fig. 14).

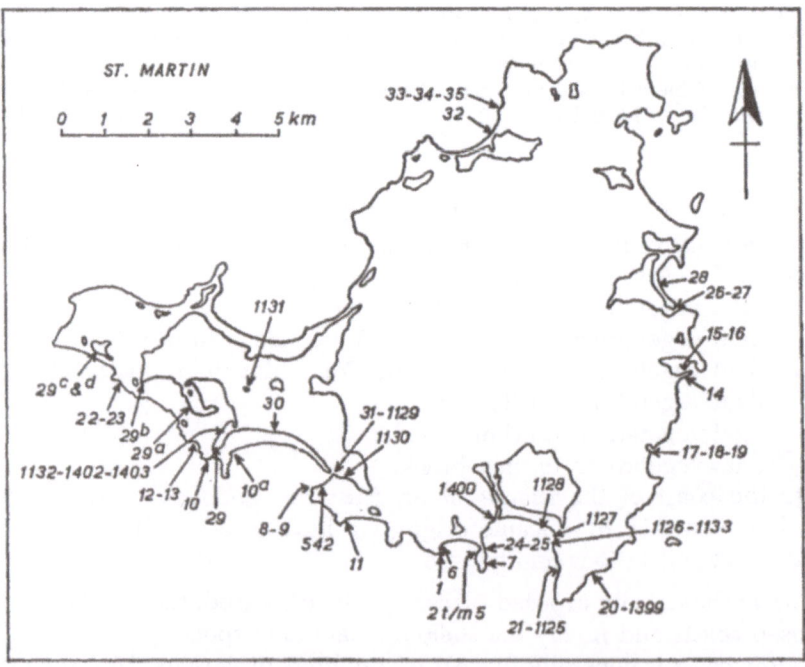

Fig. 10. Sketch map of St. Martin, with collecting numbers.

6. O y s t e r P o n d, lagoon, 12.V.1958.
Little swell; littoral and sublittoral region. — Collecting number: St. Martin 15 (Fig. 15).
7. O y s t e r P o n d, lagoon, 12.V.1958.
Free floating algae in front of coast of coral limestone. — Collecting number: St. Martin 16.

1. G u a n a B a y

Near Guana Bay (St. M. 17) the coast consists of coral limestone over a length of 150—200 m, with an outcrop of the Point Blanche formation in the middle (Pl. VIII b). As the rocks are only slightly higher than water level, they are regularly washed by the waves, up to 2—3 m (Fig. 11). The upper surface of the rocks is rough; particularly in the lower part very sharp ridges are found. Where the land vegetation begins, the surface of the rocks is rather smooth.

Several distinct, vari-coloured zones are striking (Pl. IVa). The lowest, dark-coloured zone contains pools, the water of which is continuously renewed by the waves. In these pools and on the ridges seaweeds are common. The algae causing the dark colour of this zone, reach to about 2 m above sea-level.

Above the dark-coloured zone the rocks are light-coloured. In this zone many specimens of *Nerita tessellata* are found: smaller specimens in places constantly washed by the waves, larger ones more towards land. In the light-coloured zone tufts of algae are still growing; Chitons reach about the same level as these patches of algae.

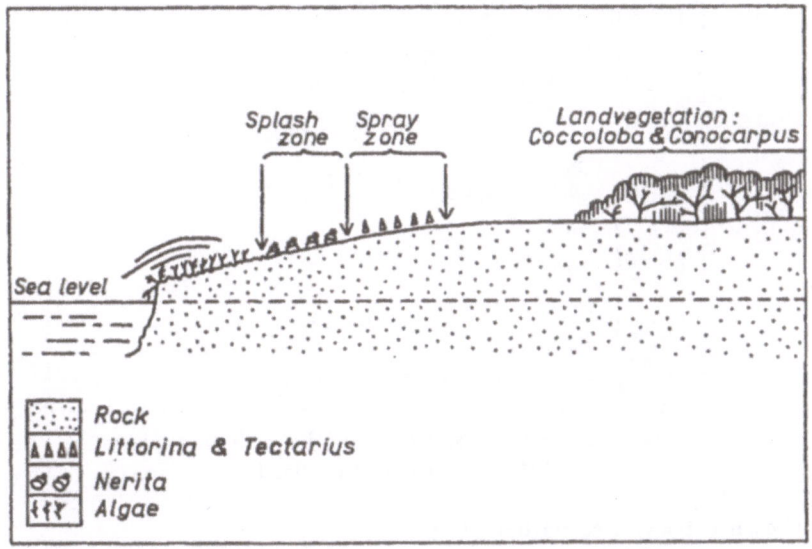

Fig. 11. St. Martin, Guana Bay: zonation on coral limestone.

Still higher, in the spray zone, where the rocks are again dark-coloured, *Littorina ziczac* and *Tectarius tuberculatus* are numerous.

Above the spray zone a land vegetation is found with many *Conocarpus erecta* and *Coccoloba uvifera* (littoral woodland, Stoffers 1956).

It is clear that algal growth is most abundant at lower levels. At the rims of the pools, under exposed conditions, *Sargassum platycarpum*, *Turbinaria turbinata* and *Laurencia papillosa* are abundant. The pools contain *Avrainvillea rawsoni*, *Halimeda opuntia*, *Rhipocephalus phoenix*, *Valonia ocellata*, *Dictyota dentata*, *Stypopodium zonale* and *Amphiroa fragilissima*. *Sargassum platycarpum* was also observed (partly with *Lophocladia trichoclados* as an epiphyte). In holes of the rock at this level the short-spined urchin *Echinometra lucunter* is abundant. *Laurencia papillosa* is also numerous in the short moss-like vegetation, covering the rock surface. Many species are growing intermingled, amongst them: *Cladophoropsis membranacea* (also forming small cushions high in the algal zone), *Dilophus guineensis* and *Padina sanctae-crucis* (both species in rather exposed places), *Centroceras clavulatum, Ceramium byssoideum* (mostly as an epiphyte on large algae, just as *Herposiphonia tenella*), *Hypnea musciformis, Jania adhaerens, Laurencia microcladia, Lophosiphonia cristata, Polysiphonia ferulacea* and *Polysiphonia* cf. *howei*.

The bluegreens *Hydrocoleum lyngbyaceum* and *Spirulina subsalsa* are common high in the algal zone.

To the south the coast of coral limestone passes into a sandy beach. On a few boulders, less exposed but continuously washed by waves, loaded with whirling sand particles, *Chondria tenuissima* and *Digenea simplex* are abundant.

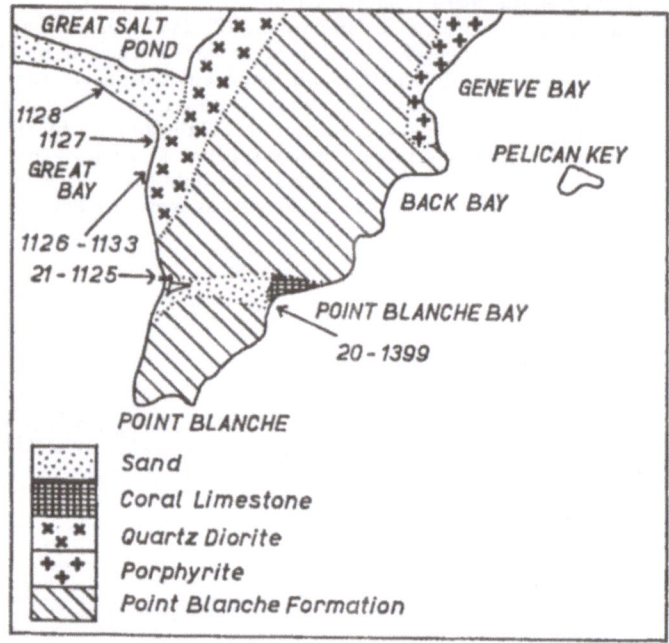

Fig. 12. St. Martin: situation of the collecting places in the S.E. part of the island.

2. Guana Bay, free floating algae

Due to the heavy surf information on the sublittoral region could not be obtained.

In front of the sandy beach many detached algae are floating at the surface (St. M. 19), mainly *Sargassum natans, Bryothamnion triquetrum, Dictyota dentata, Dilophus alternans* and *Dictyopteris plagiogramma*. Other species are *Enantiocladia duperreyi, Asparagopsis taxiformis, Lophocladia trichoclados, Codium isthmocladum* and *Stypopodium zonale*.

3. Point Blanche Bay

In Point Blanche Bay there is a boulder beach, bounded to the north by coral limestone (St. M. 20); to the south the beach passes into a steep cliff, consisting of sediments of the Point Blanche formation. The coral limestone is often very exposed.

Observations were chiefly restricted to the coral limestone north of the beach (Pl. IVb) running parallel to the prevailing winds. The rock is distinctly stratified and dips 45°–60°. Several differently coloured zones may be distinguished (Fig. 13).

Immediately above the algae, also continuously washed by the waves, a dark yellowish coloured zone of about 30 cm width may be observed. Chitons and *Nerita tessellata* are abundant; the smaller specimens of *Nerita* are found at a lower level between the algae.

Three other zones are present, each about 30 cm wide, light yellow, light grey and dark grey in colour respectively. The light yellow and light grey zones are rich in *Littorina* and *Tectarius*; both species mainly occur in horizontal parts where some water is present (spray zone). However, the small snails were also seen on completely dry parts of the rock. Above the dark grey zone, about 150–200 cm above sea-level, the rock is greyish-white and not very rough.

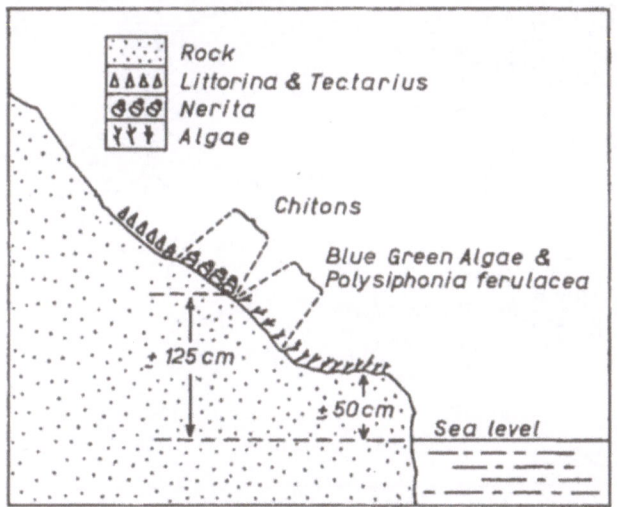

Fig. 13. St. Martin, Point Blanche Bay: zonation on coral limestone.

Just below the Chiton-zone the rocks are grown with *Polysiphonia ferulacea* and bluegreen algae; the greatest development of algae, however, is found below this zone. Abundant are *Sargassum* spec. and *Laurencia microcladia*; *Turbinaria* is common. Just below the water level we find crustlike Lithothamnia; together with *Pocockiella* they form a mosaic, covering the surface of the rock nearly completely. About at the same level as *Sargassum*, also *Lithothamnion corneum* is found. *Padina sanctae-crucis* and *Dictyota* spec. appear to be sparse; the latter grows somewhat below *Sargassum*. An almost complete absence of *Laurencia papillosa* may be noted.

On the pebble beach much *Sargassum natans* and *Halimeda opuntia* is washed ashore.

On the rocks of the Point Blanche formation at the southern end of the Bay remarkably few algae are present. Chitons were not observed at all.

4. Little Bay

At the eastern side of Little Bay, well sheltered against wave action, a small part of the coast is formed by coral limestone (St. M. 5; Fig. 14), up to 50–100 cm above sea-level. The depth of the water in front of the rock varies from 30 to about 125 cm and the bottom consists of clean sand (Pl. IIIb). Here, several large boulders are continuously washed by waves, loaded with whirling sand. Nevertheless, these

45

rocks have a dense algal vegetation of *Padina sanctae-crucis*, *Bryothamnion triquetrum*, *Chondria tenuissima*, *Laurencia papillosa* and *Sargassum vulgare*.

The coral limestone has a small niche, in which many algae are present. Near the water level *Dictyota ciliolata*, *Dilophus guineensis*, *Cladophoropsis membranacea* and *Cladophora luteola* are observed. *Laurencia obtusa* (0—50 cm deep), *Anadyomene stellata* (near to the water level), *Falkenbergia hillebrandii* (at 0—30 cm) together with its gametophyte *Asparagopsis taxiformis* (20 cm deep), *Wrangelia argus* and *Bryopsis pennata* have been also collected.

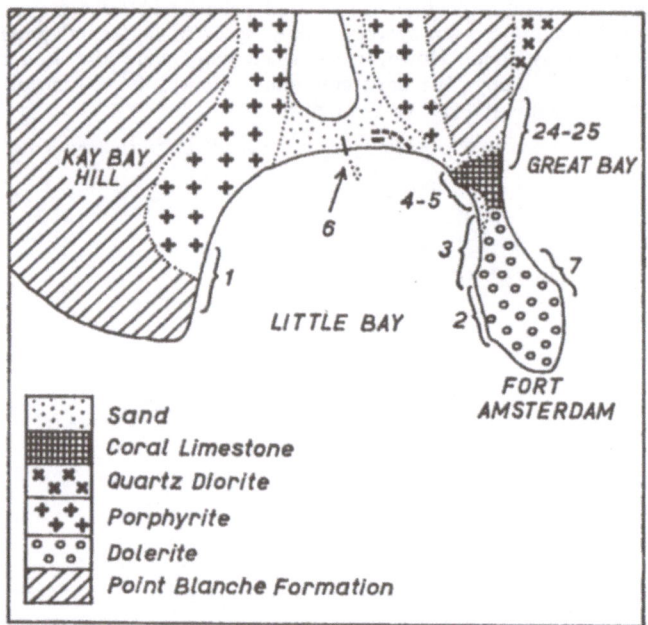

Fig. 14. St. Martin, Little Bay and Great Bay: situation of sampling spots.

A single specimen of *Halimeda opuntia* was found at a depth of about 100 cm. The segments of the thallus appear to be remarkably thin in this dark location.

Against the vertical parts of the rock, above sea-level and shadowed, large patches of *Lophosiphonia cristata* grow, together with a dark-coloured alga (probably the bluegreen *Dichotrix fucicola*).

5. Little Bay, rocks along the shore

Somewhat south of the preceding part of the coast several boulders of limestone are found on a sandy bottom, at a depth of about 100 cm (St. M. 4). These rocks are also continuously wetted by the waves and show an algal vegetation of *Chondria tenuissima*, *Padina sanctae-crucis*, *Laurencia papillosa* and *Laurencia poitei*. Barnacles are present at water level.

6. Oyster Pond, lagoon

The coast of the lagoon of Oyster Pond (St. M. 15) partly consists of coral limestone (Fig. 15). The easternmost part of the spit of land that separates the lagoon from the sea emerges about 30—50 cm above high water mark. The water in the lagoon has a depth of about 50 cm and the bottom is sandy with stones. On these rocks small barnacles were found, washed by low waves.

On the limestone striking, continuous strips of *Bostrychia tenella* (about 10–20 cm above the water level), and *Polysiphonia howei* (about 10 cm above) have developed. On the roots of mangroves both species are also abundant. They grow only in places without direct sunlight.

Just below the water there are several specimens of *Chaetomorpha media, Dictyosphaeria vanbosseae* and *Gracilaria cervicornis. Ulva lactuca* is abundant. In a few places, at about 20 cm, *Pterocladia pinnata* forms large patches on the rocks. Of *Codium taylori,* one single plant was found. Poorly developed *Sargassum polyceratium* is also present.

On limestone rocks in front of the shore, at a depth of about 30 cm, *Acanthophora spicifera, Galaxaura squalida* and *Dictyota* spec. occur, together with abundant *Laurencia papillosa. Chaetomorpha crassa* and *C. linum* were common on all large algae, mostly strongly intermingled. In front of the coral limestone the bottom is sandy, and has a dense growth of *Thalassia*.

7. O y s t e r P o n d, lagoon; free floating algae

In the lagoon many free floating algae were found (St. M. 16): *Bryothamnion seaforthii, Pocockiella variegata, Cladophora fascicularis, Galaxaura marginata, Dilophus alternans*. Also present are *Dictyopteris delicatula, Ceramium fastigiatum, Centroceras clavulatum, Chaetomorpha crassa* and *Dictyota* spec. Part of these algae are epiphytes on larger species. This sample probably contains algae floated from the sea into the lagoon.

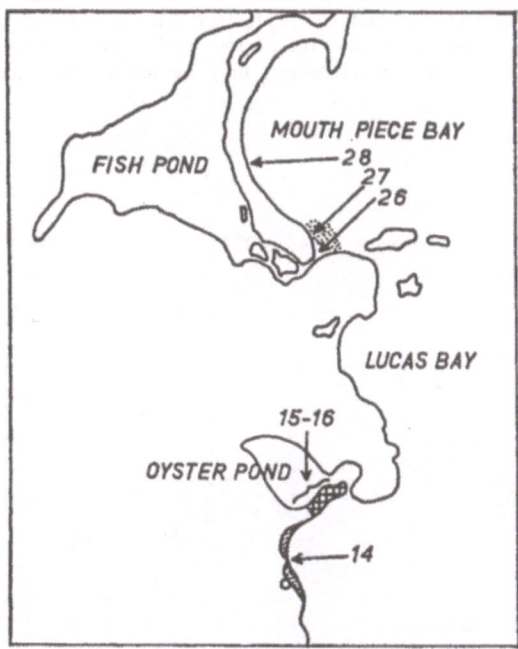

Fig. 15. St. Martin: situation of the collecting places at Mouth Piece Bay and Oysterpond.

B. 2. Algal vegetations on rocks of the Low Lands formation
    (*St. Martin*)

The rocks of the Low Lands formation are found exclusively in the western part of St. Martin (Fig. 16).

Legenda:

1. Cupecoy Bay, 16.V.1958.
Rocks W. of Bay, partly forming a terrace of 1—2 m width. Heavy swell; littoral region and upper part of sublittoral region. — Collecting number: St. Martin 23.

2. Maho Bay, 9.V.1958.
Flat rock at the boundary of sandy beach and cliff, about 50 cm above sea-level. Heavy swell; littoral and upper part of sublittoral region. — Collecting number: St. Martin 12 (Pl. VIb).

3. Maho Bay, 9.V.1958.
Steep cliff, in front of it, at a depth of about 3—4 m, a flat rock. Sublittoral region. — Collecting number: St. Martin 13.

4. Burgeux Bay, 7.V.1958.
Rocky coast W. of bay and utmost western part of bay. Heavy swell; littoral region and upper part of sublittoral region. — Collecting number: St. Martin 10 (Pl. VIa).

5. Simson Bay, eastern part, 6.V.1958.
Slightly exposed; littoral region and upper part of sublittoral region. — Collecting number: St. Martin 8 (Pl. Vb).

In the extreme west of St. Martin, near Long Bay there is a long sandy beach; the seabottom in front of the beach is also sandy over a large area. East of Long Bay the rocky coast of the Low Lands formation begins, and continues to the east with an interruption at Simson Bay as far as Cole Bay (Fig. 2). The rocky parts, protruding into the sea between the bays, are rather low and have a Karren appearance. The bays generally have a sandy beach which may be made very steep by strong wave

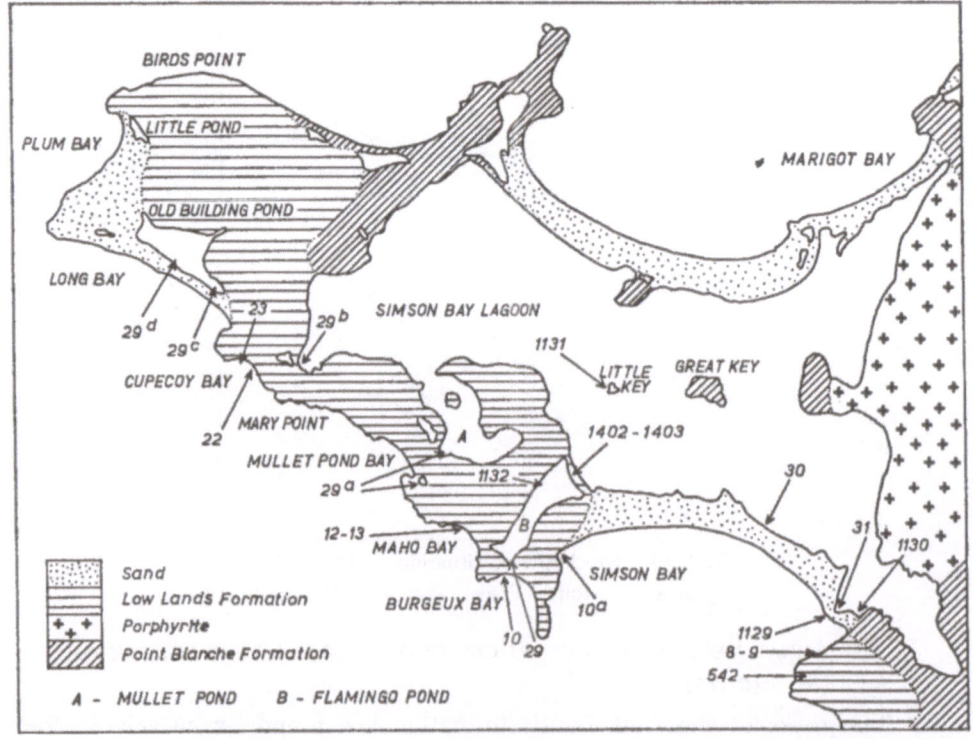

Fig. 16. St. Martin: situation of the sampling spots in the western part of the island.

action to 1—1½ m above sea-level. In other bays a steep cliff is present, and the rocks of the Low Lands formation rise vertically to 7—10 m. In several places before these cliffs there is a narrow sandy beach, e.g. in Cupecoy Bay, the most western part visited in the Low Lands. Beachrock plates may occur.

At Mary Point and Mullet Pond Bay (Pl. II) a narrow sandy beach is found in front of a steep cliff, which only in three places is in direct contact with the sea. Below water level a number of flat rocks is found, covered by a thin layer of moving sand.

Near Mary Point a number of rocks rests in the seabottom, with their upper surface at about sea-level. These rocks, which are washed by waves loaded with whirling sand, have an abundance of *Chondria tenuissima*. On the rocky coast at Mary Point, a number of clearly different coloured zones is observed. Above the zone with algal growth a light-coloured band and a dark zone are present. The three zones have a width of about 1 m each. The coast in this place has a height of at least 7 m and is inaccessible.

Immediately west of Mullet Pond Bay numerous rocks are found along the coastal line. The rocks projecting into the sea have a dense vegetation of *Sargassum* spec. and *Turbinaria turbinata*, the rocks lying more towards land are washed by waves loaded with sand, and have a heavy growth of *Chondria tenuissima*.

West of Maho Bay the rock rises steeply from the sea to about 3—4 m. At the boundary of rocky shore and sandy beach a flat rock is found, about 50 cm above sea-level. In front of the cliff, at a depth of 3—4 m, again a flat rock is present. Distinct niches are formed (Pl. VIb).

Also Burgeux Bay is bordered on both sides by a steep cliff. The sandy beach with large plates of beachrock is subjected to heavy wave action. Immediately in front of the rocky coast corals are present.

In Simson Bay there is a beautiful faintly curved sandy shore wall, separating Simson Bay from Simson Bay Lagoon. On the sandy bottom of Simson Bay a dense vegetation of seagrasses can be seen and in the western part of the bay a thick layer of detached leaves is cast ashore. These leaves reach a height of half a meter and a width of 1—2 m; they are almost exclusively leaves of *Syringodium* (St. M. 10a). Large quantities of these leaves were floating in this part of Simson Bay.

In the eastern part of Simson Bay the coast is formed by large boulders, partly of the Point Blanche formation, partly of the Low Lands formation (Pl. Vb).

Near Lay Bay (Pl. Ib) the coast projects again into the sea forming many small bays with sandy beaches.

In Cole Bay, a transition of the coast into rocks of the Point Blanche formation is observed.

1. Cupecoy Bay

In Cupecoy Bay (St. M. 23) the swell is heavy. Algae are best developed on a terrace, 1—2 m wide, lying just above mean sea-level and constantly washed by the waves, loaded with sand. The rocky coast shows a marked zonation, as was seen at Mary Point, with an algal zone, a light-coloured and a dark zone above each other. Chitons are absent; *Littorina* and *Tectarius* appear to be numerous in the dark zone.

The short moss-like developed vegetation of the terrace appears to be very dense and consists of many intermingled species. Most is formed by *Jania adhaerens* (abundant), *Polysiphonia ferulacea*, *Dilophus guineensis*, *Padina sanctae-crucis* and *P. vickersiae*. Also present are *Chondria tenuissima*, *Laurencia papillosa*, *L. microcladia*, *Herposiphonia tenella* (epiphytically on several large species), *Spyridia aculeata* (not common) and *Dichothrix fucicola*. *Padina* and *Chondria* are abundant on horizontal parts of the rock; *Laurencia microcladia* particularly in strongly exposed places.

2. Maho Bay

The sandy beach of Maho Bay (St. M. 12) is bordered to the west by rocks rising

49

vertically from the sea as high as 3–4 m. Between the sandy beach and the rocky coast a flat rock, about 50 cm above sea-level, is found, regularly wetted by the waves. The coast is not very rich in species. The swell of the waves is heavy: Chitons are seen 50–100 cm above sea-level; *Tectarius* and *Littorina* are present up to 3 m.

At the foot of the flat rock *Bryothamnion triquetrum* is abundant, washed by sand-loaded waves. On the upper side, several square meters are covered with *Chondria tenuissima*, *Polysiphonia ferulacea* and *Padina sanctae-crucis*. The algal growth is partly dense and moss-like, consisting of the following intermingled species: *Padina sanctae-crucis*, *Polysiphonia ferulacea*, *Dilophus guineensis*, *Spyridia aculeata*, *Cladophoropsis membranacea*, *Cladophora fuliginosa*, *Laurencia microcladia* and *L. papillosa* (the last mentioned species inter alia with *Dipterosiphonia dendritica* as an epiphyte).

On the rocky coast, and at water level, grow *Sargassum* cf. *filipendula* and *Turbinaria*, and immediately below several specimens of *Ochtodes filiformis*. In fissures and small pools both *Cladophoropsis membranacea* and *Cladophora fuliginosa* are rather common. A crustlike Corallinaceae and *Pocockiella* cover the rocks.

3. Maho Bay, sublittoral region

The rocky shore of Maho Bay appears to be rather steep also below the water level (St. M. 13). At a depth of 3 to 4 m a flat rock is found, about 10 m wide, covered only by a thin layer of sand with many specimens of seafans and a number of algae. The water appeared to be in heavy motion down to this depth.

The vertical rock does not show many algae, but harbors several animals, e.g. the stinging coral, *Millepora alcicornis* and the urchin *Diadema antillarum*. Several specimens of *Halimeda opuntia* were collected, together with *Wrangelia argus* and *Laurencia obtusa*. *Dilophus alternans* grows from 0–150 cm, densely covered with *Melobesia farinosa*; *Ochtodes filiformis* is rather abundant from 0–100 cm.

On the flat rock, next to the luxurious growth of seafans, several plants of *Galaxaura rugosa* occur at a depth of 4 m, partly covered with numerous epiphytes, i.a. *Spermothamnion investiens*, *Diplochaete solitaria*, *Asterocytis ramosa*, *Griffithsia tenuis*, *G.* cf. *globulifera*, *Champia parvula*, *Ceramium byssoideum* and *Hypnea* spec. In the layer of sand several specimens of *Rhipocephalus phoenix*, *Udotea sublittoralis* and *Udotea flabellum* are present. *Halimeda discoidea* and *Amphiroa hancockii* grow at the same depth.

4. Burgeux Bay

The rocky shore in the western part of Burgeux Bay (St. M. 10) is 1–2 m high. The swell is rather heavy and algal growth is only possible to about 50 cm above sea-level (to the Chiton-zone). *Littorina* and *Tectarius* are found in the spray zone. In front of the coast, many corals are growing, inter alia *Acropora palmata*.

The steep rocky shore shows many algae. A bluegreen alga forms smooth, black pellets, high on the shore. *Padina sanctae-crucis* occurs in large numbers on several flat rocks; *Sargassum polyceratium*, *S. platycarpum* and *S. rigidulum* grow at a somewhat lower level. *Sargassum polyceratium* prefers exposed places, generally together with *Turbinaria*, which is also abundant; *S. platycarpum* is mainly found in pools.

The rocks are in part covered with a close, moss-like algal vegetation, containing much sand. Several species grow intermingled, e.g. *Padina sanctae-crucis*, *Cladophoropsis membranacea*, *Dictyopteris delicatula* (partly epiphytically on *Sargassum* spp. and *Padina sanctae-crucis*), *Laurencia papillosa* (common), *L. microcladia*, *Polysiphonia ferulacea* (abundant), *Herposiphonia* cf. *tenella* (partly as an epiphyte on *Laurencia papillosa*), *Jania adhaerens* (rather numerous), *Lophosiphonia cristata*, *Spyridia aculeata* and *S. filamentosa*. Of *Colpomenia sinuosa* only a few plants are found. A few specimens of *Laurencia obtusa* were observed at a somewhat lower level than *L. papillosa*. Of *Chaetomorpha media* two plants were gathered at the low water line.

50

In very exposed places, below the zone with *Sargassum*, *Lithothamnion* cf. *corneum* may be found.

Below *Sargassum*, *Padina sanctae-crucis* and *Chondria tenuissima* are present: *Dictyota ciliolata*, *D. dentata*, *Dilophus alternans* and *D. guineensis*.

5. Simson Bay

In Simson Bay (St. M. 8; Pl. Vb), next to rocks of the Low Lands formation, boulders of the Point Blanche formation are found, but no significant differences in zonation are observed. On the other hand, depending on the exposure to the wind, obvious differences exist. The most north-eastern part of the bay has a rather steep coast, with large boulders on a sandy bottom.

Barnacles, small specimens of *Nerita* and Chitons are seen 5—20 cm above sea-level; small mussels in abundance 0—10 cm above. *Sargassum platycarpum* and also *Padina sanctae-crucis*, *Dilophus guineensis*, *Laurencia papillosa* and *L. obtusa* grow near the water level; *Cladophoropsis membranacea* somewhat higher. In cavities of the rock numerous specimens of the urchin *Echinometra* are found.

Close to the shore is a small island (Fig. 16, about the arrow of St. M. 542). On the rocks opposite this island, the influence of the wave action is clearly noticeable. Here the Chitons are 20—45 cm above sea-level; somewhat higher *Littorina* and *Tectarius* are found. *Nerita tessellata* occurs below the chitons. Somewhere near the Chiton-zone, a light-coloured band of about 40—50 cm wide may be seen around the boulders, probably formed by dead Lithothamnia. The same is observed on doleritic rocks near Fort Amsterdam (Fig. 17) and on Saba.

The bottom underlying the rocks mainly consists of sand; the water has a depth of 50—100 cm. *Bryothamnion triquetrum*, *Digenea simplex*, and *Padina sanctae-crucis* are found at sea-level; *Chondria tenuissima* from about 10 cm above to 30 cm below.

On the boulders, opposite the small island, *Sargassum platycarpum* is very abundant, mostly together with *Turbinaria*. On these rocks and close to the water level many species grow together in a dense moss-like vegetation: *Anadyomene stellata*, *Caulerpa racemosa*, *Cladophoropsis membranacea*, *Dictyosphaeria cavernosa*, *Dictyota ciliolata*, *D. dentata*, *D. dichotoma*, *D. divaricata*, *Dilophus alternans*, *D. guineensis*, *Padina sanctae-crucis*, *Jania adhaerens*, *Laurencia papillosa*, *Lophosiphonia cristata*, *Lithothamnion* spec., *Polysiphonia ferulacea*, etc. Somewhat deeper many species grow attached to the boulders: *Acetabularia crenulata*, *Avrainvillea rawsoni*, *Neomeris annulata*, *Galaxaura rugosa*, *G. subverticillata*, *Gracilaria ferox*, *G. debilis*, *Laurencia obtusa*, *Liagora pedicellata*, *L. valida*, *Lithothamnion* cf. *occidentale*.

In the sandy bottom between the boulders also a number of species grows. Small patches of *Thalassia* are formed, often projecting about 30 cm above the surrounding unstable sandy bottom (cf. IV C. 1).

B. 3. ALGAL VEGETATIONS ON ROCKS OF THE POINT BLANCHE FORMATION (*St. Martin*)

Next to the peninsula of Point Blanche, rocks of this formation are found in different places along the coast of St. Martin, usually in large boulders with a diameter of more than 1 m. The smooth surface of cleavage planes in this layered rock is remarkable. The sides of the boulders show small ridges which provide good possibilities for the attachment of algae.

Legenda:

1. Oyster Pond, bay, 12.V.1958.
Protruding rocks with a sandy beach on both sides. Heavy swell; littoral region and upper part of the sublittoral region. — Collecting number: St. Martin 14 (Pl. VIIIa; Fig. 15).

2. G u a n a B a y, 13.V.1958.
Projecting rocks between coast of coral limestone. Heavy swell; littoral and upper part of the sublittoral region. — Collecting number: St. Martin 18 (Pl. VIIIb; Fig. 18).
3. L i t t l e B a y, eastern slope of Kay Bay Hill, 28.IV.1958.
Rather heavy swell; littoral region and upper part of sublittoral region. — Collecting number: St. Martin 1 (Pl. VIIa; Fig. 14).
4. G r e a t B a y, 14.V.1958.
Rocks at both sides of small landing-stage. Slightly exposed, littoral and sublittoral region. — Collecting number: St. Martin 21 (Pl. VIIb; Fig. 12).

1. Oyster Pond

In the sandy bay of Oyster Pond (St. M. 14) a number of boulders of the Point Blanche formation projects into the sea. The most outward rocks have the best developed algal vegetation; more landward rocks are less covered with algae, because the water is heavily loaded with sand. The rocks in Oyster Pond have remarkably rounded erosional forms. Hardly any Chiton is seen and also *Littorina* and *Tectarius* appear to be scarce; *Pocockiella* and the crustforming Lithothamnia, which in many places cover large areas of the surface of the rock, are not observed.

On the most outwardly situated, exposed rocks rather a lot of algae are present: *Sargassum platycarpum* and *S. polyceratium* are common; of *Turbinaria* a few small specimens are present. *Laurencia microcladia* and *Laurencia* spec. occur in exposed places. Abundant are *Grateloupia filicina, Polysiphonia ferulacea* and *Gracilaria ferox*. Partly, the algae form a short dense vegetation, with i.a. *Dictyota ciliolata, Dilophus guineensis, Cladophoropsis membranacea, Laurencia papillosa, L.* cf. *obtusa, Centroceras clavulatum, Jania* spec. and *Hypnea musciformis*.

On the more inwardly situated rocks, washed by waves loaded with sand, *Bryothamnion triquetrum* (just below the water-line) and *Chondria tenuissima* (abundant) are present. A number of smaller species may be found between these algae. In the rather quiet water between the boulders were observed: *Halimeda opuntia, Cladophoropsis membranacea, Dictyota* spp., *Chaetomorpha media* and *Enteromorpha flexuosa*. Many epiphytes grow on *Digenea simplex: Jania rubens, J. adhaerens, Corallina cubensis, Lophosiphonia cristata, Hypnea spinella, Ceramium* spec. and *Herposiphonia tenella*.

2. Guana Bay

Near Guana Bay (St. M. 18) a number of rocks of the Point Blanche formation projects between rocks of coral limestone. On large boulders (at least 2 m or more high) some zonation is visible. The lowest parts of the rock are dark-coloured and densely covered with algae. At a higher level, where the waves constantly wash the rocks, a light-coloured zone is seen, and still higher another dark zone.

On exposed rocks the following algae are abundant, partly intermingled: *Sargassum platycarpum, Turbinaria turbinata, Laurencia papillosa, L. microcladia, Polysiphonia ferulacea, Padina sanctae-crucis, Hypnea musciformis, Lophosiphonia cristata, Dictyota dentata* and *Dilophus guineensis*. In quiet water, between and behind the rocks, and mostly also constantly submerged, a number of other species is present: *Caulerpa cupressoides, Avrainvillea rawsoni* (a few specimens), *Neomeris annulata, Halimeda opuntia, Amphiroa fragilissima, Udotea conglutinata,* and *Cladophoropsis membranacea*.

3. Little Bay

In the southwestern part of Little Bay, near the eastern slope of Kay Bay Hill (St. M. 1) the swell is rather heavy. The diameter of the (partly porphyritic) boulders is variable.

Chiton is scarce about 40 cm above low water level. The upper limit of algal growth coincides with that of chiton, *Padina sanctae-crucis* in many cases goes to about the same height and is only found on parts of the rock sloping 45° or less.

52

The width of the *Padina* zone appears to be related to the dip of the rocky surface, but in general 30—40 cm above the water level large numbers of this species are present. *Sargassum* spec. usually grows at a somewhat lower level than *Padina* but also on nearly horizontal parts of the rock. *Turbinaria* again grows somewhat lower and on nearly horizontal rocks. On parts of the rock steeper than 45° (up to 90°) quite a number of species is present: e.g. *Laurencia papillosa* and *L. microcladia* just below *Padina sanctae-crucis*.

On nearly constantly submerged parts of the rock, a crustlike lime-incrustating red alga is found, at the sheltered side completely covering the surface of the rock, including limpets and barnacles.

At the exposed side the Lithothamnia also occur to about the water level, but cover the rock incompletely on account of many other moss-like developed algae. Of frequent occurrence is also *Pocockiella variegata;* in quiet water this species is not able to withstand the competition with the Lithothamnia, but it prefers the same conditions. *Pocockiella* grows to heights where the rocks are wetted by waves and reaches certainly 20 cm downward. At the exposed outer side of the rock it generally forms a mosaic with the Lithothamnia and other, mostly moss-like developed algae.

Between the boulders the water is usually much more quiet. Here, as described above, Lithothamnia are well developed. Also several corals are present, especially *Millepora alcicornis*, and flat colonies of *Eusmilia fastigiata*, with in some places also *Porites porites*. Wherever corals are found Lithothamnia are absent. Barnacles are seen 0—15 cm above the water level. *Nerita tessellata* is observed from about 5 cm above the water to a depth of about 20 cm. *Littorina* and *Tectarius* occur above *Nerita* and may live to 150 cm above water level.

In the quiet water between the boulders, between the corals and Lithothamnia there are also many urchins, such as *Echinometra* and *Diadema*. In places many small mussels are present, covering the bottom completely; *Purpura* is abundant too. Next to these animals several algae are encountered. *Amphiroa fragilissima* forms large cushions in several places; *Halimeda opuntia* is represented by several large plants; *Chaetomorpha media* grows at about low water level; *Struvea anastomosans* is rather abundant in dark places. *Valonia ventricosa, Dictyosphaeria vanbosseae, Wrangelia argus, Cladophoropsis membranacea, Neomeris annulata* and *Ectocarpus breviarticulatus* have also been observed. In exposed places other species are also present, such as *Dictyota dentata, Dilophus guineensis, Ochtodes filiformis, Hypnea musciformis, Polysiphonia ferulacea, Asparagopsis taxiformis, Spyridia aculeata, Centroceras clavulatum, Valonia ocellata, Jania adhaerens* and *Laurencia gemmifera*. Several of the larger species carry epiphytes: *Diplochaete solitaria, Herposiphonia pectenveneris, H. tenella, H. secunda, Melobesia farinosa, Entocladia* spec., *Acrochaetium* spec., *Ceramium floridanum, Falkenbergia hillebrandii. Polysiphonia howei* is found in a few places, probably only wetted by spray.

4. G r e a t B a y, near small landing-stage

Finally, samples have also been taken on rocks of the Point Blanche formation in the eastern part of Great Bay (St. M. 21) near a small landingstage.

Approximately at sea-level *Padina sanctae-crucis* is abundant, together with many small mussels and barnacles. *Nerita* is found to 10 cm above *Padina*, where it is constantly washed by the waves. The rocks are only found close to the shore; outwards the bottom consists of clean sand without seagrasses or algae, down to 3 m. The rocks are barren in the sublittoral region, even coral is not present, but several *Diadema* have been observed.

The following algae were found: *Turbinaria* (several specimens), *Sargassum platycarpum* (at a depth of about 50 cm), *Sargassum vulgare* (higher than the preceding species), *Cladophoropsis membranacea, Cladophora fuliginosa, Dictyota dentata, Dilophus alternans, D. guineensis* (the last mentioned three species at about 50 cm), *Galaxaura squalida, Halimeda opuntia* and *Neomeris annulata*.

B. 4. ALGAL VEGETATIONS ON DOLERITIC ROCKS (*St. Martin*)

Near the peninsula of Fort Amsterdam the greater part of the coast (Fig. 14) is formed by angular boulders of dolerite with a smooth surface; a small part appears as a steep cliff.

Legenda:

1. Great Bay, 2.V.1958.
Steep cliff at the eastern side of the peninsula with Fort Amsterdam. Heavy swell; littoral region and upper part of sublittoral region. – Collecting number: St. Martin 7 (Pl. IIIa; Fig. 17).
2. Great Bay, 22.V.1958.
Boulder coast. Rather heavy swell; littoral region and upper part of the sublittoral region. – Collecting number: St. Martin 25 (Pl. III).
3. Great Bay, 22.V.1958.
Boulders in front of the coast, partly resting in a sandy bottom. Rather heavy swell; sublittoral region. – Collecting number: St. Martin 24.
4. Little Bay, east of sandy beach, 29.IV.1958.
Boulder coast. Slightly exposed; littoral region and upper part of sublittoral region. – Collecting number: St. Martin 2 (Pl. IIIb).
5. Little Bay, 30.IV.1958.
Boulders in front of doleritic coast in eastern part of the bay. Slightly exposed habitat; sublittoral region. – Collecting number: St. Martin 3 (Pl. IIIb).

1. Great Bay

In front of the doleritic rocks of a steep cliff (Pl. IIIa; St. M. 7) we find a number of large, angular, but smooth boulders. The waves run high on the coast and all rocks are continuously washed.

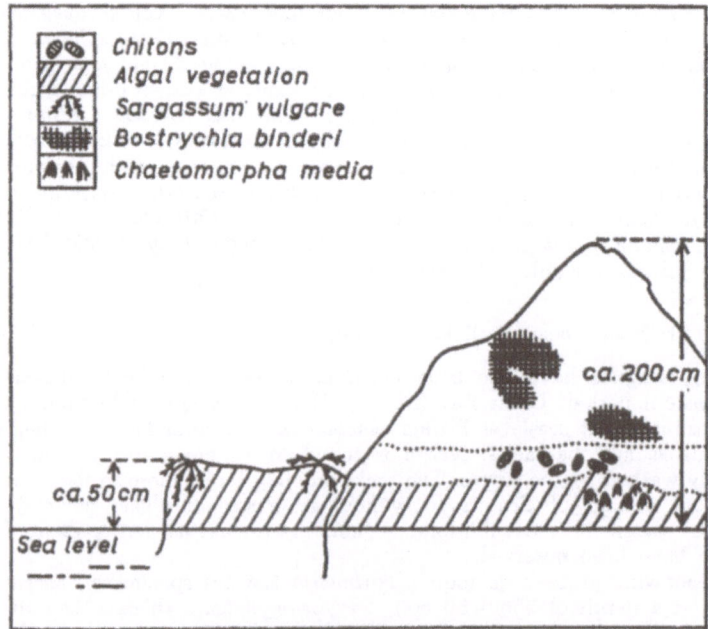

Fig. 17. ST. MARTIN, Great Bay: algal vegetation on dolerite rock, east of the peninsula with Fort Amsterdam.

Sargassum vulgare grows on flat parts of the rock, about 20 cm above high water level; *Padina sanctae-crucis* immediately above *Sargassum*. *Laurencia microcladia*, on the other hand, grows, together with other *Laurencia* species, preferably on vertical rock. *Laurencia papillosa* grows at about the same level as *Sargassum*; *L. obtusa* is numerous in the upper part of the sublittoral, and *L. cf. obtusa* forms thick cushions in several rather exposed places. *Pocockiella* is abundant in the higher places. Also several crustlike Lithothamnia are present. *Chaetomorpha media* indicates the low water line. Barnacles are fairly scarce and occur 20 cm or more above the seaweeds. Chiton is found in a zone of about 20 cm wide above the algae. In one place Chitons occur up to 200 cm above sea-level; the waves reach to this height and the rock does not receive any direct sunlight. At the same place *Bostrychia binderi* and *Polysiphonia howei* occur, smaller in size than on *Rhizophora* roots.

In many places a light-coloured zone may be observed at about the same height as the Chiton zone, and about 20 cm wide: possibly dead crustlike Lithothamnia (Fig. 17). *Littorina* and *Tectarius* grow at a higher level, on dry rocks.

Two species of urchins, viz. *Lytechinus* and *Echinometra*, live in the shallow water between the boulders. In the rather shallow water in front of the shore much coral is present, i.a. *Acropora palmata*, but algae are practically absent.

In small pools between the rocks, where the water is constantly renewed by the waves, the following algae are found: *Caulerpa racemosa, C. sertularioides, Halimeda opuntia, Acanthophora spicifera* and *Dictyosphaeria cavernosa*; in the splash zone: *Polysiphonia ferulacea* (abundant), *Grateloupia cuneifolia* (rather abundant), *Cladophoropsis membranacea* (common), *Pterocladia pinnata, Grateloupia filicina, Lophosiphonia cristata, Pocockiella variegata*.

2. Great Bay, boulders in shore line

Somewhat north of the locality described above, behind the annexes of Little Bay Hotel, the coast is formed by strongly grounded doleritic rocks (St. M. 25). The algal vegetation is rather poor: *Chaetomorpha media* forms 90% of the vegetation, other species grow in small tufts between its rhizoids. Of *Laurencia obtusa* several specimens have been collected, and *L. intricata* is also present. Several small plants of *Ulva* cf. *lactuca* have been found on a rather high level. Other accompanying species of *Chaetomorpha media* are *Struvea anastomosans, Hypnea musciformis, Centroceras clavulatum, Giffordia mitchellae*, and *Ectocarpus confervoides*.

3. Great Bay, sublittoral region

The western shore of Great Bay has a rather strong swell. On May 22, 1958, however, the quiet weather permitted an investigation of part of the bay to about 50 m off shore (St. M. 24).

The algal vegetation is very rich, in contrast to the vegetation on the boulders along the shore (St. M. 25). The bottom in front of the coast consists of rounded doleritic boulders, further out the bottom consists of sand. The water remains shallow for at least 10 m from the shore. The bottom slopes gradually to a depth of about 2 m at 50 m off shore.

Between the boulders many urchins are present; in shallow water mainly *Echinometra* and *Lytechinus;* at a depth of 30—50 cm *Diadema*. A lot of coral is found, e.g. *Acropora cervicornis, A. palmata*, and *Porites*, together with numerous gorgonids.

A rich growth of algae is present on the boulders. Crustlike Lithothamnia are abundant on all suitable substrates in a zone immediately above and below the low-water line, and *Pocockiella* is also present. A *Lithothamnion* species, forming thick knobs, is attached to rocks and dead coral.

On many boulders algae form a dense vegetation: *Bryopsis pennata* (one specimen at 1 m below a rock), *Chaetomorpha media* (in some places to a depth of 30 cm, but also much higher on the shore), *Cladophora* spec., *Cladophoropsis membranacea* (on *Laurencia obtusa*), *Ernodesmis verticillata* (one specimen at about 20 cm), *Struvea anastomosans* on *Valonia ventricosa* (rather abundant), *Amphiroa fragilissima* (on rocks just in front of the shore), *Centroceras clavulatum*, *Ceramium byssoideum*, *Crouania attenuata* (abundant at 20—50 cm), *Champia parvula*, *Gracilaria mammillaris* (in one place), *Hypnea spinella* (not common, on rocks immediately below the surface), *Jania adhaerens* (partly as an epiphyte on larger algae), *Laurencia obtusa*, *Melobesia farinosa* (epiphytically on *Valonia* and on *Thalassia*), *Pterocladia pinnata* (common on several rocks, 20 cm deep) and *Wrangelia argus* (several specimens in the same places as *Crouania*, at 20—50 cm).

4. Little Bay, doleritic rocks

The doleritic boulders in Little Bay (St. M. 2) vary in diameter from 30 to 100 cm. The surface is smooth and the algae could be sampled easily.

Many barnacles occur from 10 cm above to 10 cm below water level; Chitons 10—20 cm above water level. In shallow water *Millepora alcicornis* is abundant; *Acropora palmata*, *Porites* and *Eusmilia fastigiata* are represented by a few specimens only. Between the rocks, at water level, many small mussels are found.

The algal vegetation is not completely closed. Between the rocks *Echinometra* is common. The flat parts of the rock are overgrown by *Sargassum polyceratium*, *Turbinaria turbinata*, *Padina sanctae-crucis*, *P. vickersiae*, *Centroceras clavulatum*, *Laurencia papillosa* (numerous). On vertical parts of the rock *Laurencia microcladia* is particularly well developed. *Pocockiella variegata* is abundant, loosely attached to the rocks. Lithothamnia are not numerous. *Polysiphonia howei* forms a dense moss-like vegetation on a shady, vertical part of the rock, 100 cm above water level.

In the sublittoral region, and attached to the rocks, *Neomeris annulata* is rather abundant, and of *Gymnogongrus tenuis* one specimen has been collected. Rather abundant is *Falkenbergia hillebrandii*, also on coral limestone due north of this place (see St. M. 5). *Cladophoropsis membranacea* is found between the rocks, generally below water level.

5. Little Bay, sublittoral region

In this locality, very near to the fore-mentioned places (St. M. 3), animals are practically the only organisms and algae are hardly present. Numerous specimens of *Diadema* live between the rocks. The following corals are conspicuous: *Millepora alcicornis* (abundant), *Eusmillia fastigiata* and *Porites porites* (rather common), *Agaricia agaricites* (several specimens) and *Montastrea*. Between, only a few algae have been observed: *Dilophus alternans* (at 100 cm), *Wrangelia argus* and *Laurencia obtusa* (several plants), *Dasya* spec. (one small plant), *Turbinaria* (one specimen at 50 cm), *Galaxaura* spec. and several bluegreen algae.

B. 5. ALGAL VEGETATIONS ON DIORITE (*St. Martin*)

In Baie de la Grande Case the coast is partly formed by large diorite rocks. They have a diameter of 50 to 200 cm, and are seen to about 10 m off shore. More outwards the bottom consists of clean sand. The water has a depth of about 2 m over a large distance.

Legenda:
1. Baie de la Grande Case, 26.V.1958.
Slight wave action; littoral and upper part of sublittoral region. — Collecting number: St. Martin 33 (Pl. IXa; Fig. 18).
2. Baie de la Grande Case, 26.V.1958.
As preceding number; sublittoral region. — Collecting number: St. Martin 34.

1. Baie de la Grande Case

In Baie de la Grande Case (St. M. 33) on dioritic rocks numerous large barnacles are found 0—20 cm above water level; the seaweeds mostly remain below this zone. Chitons are lacking; *Littorina ziczac* and *Tectarius tuberculatus* are present in dry places; *Nerita tessellata* lives amongst the barnacles. Between the rocks numerous small mussels are present up to about water level. *Echinometra* is found in the same places.

Cladophoropsis membranacea forms small cushions from 20 cm below to 10 cm above waterlevel. *Pocockiella* is numerous up to waterlevel; the crustlike Lithothamnia mostly remain somewhat lower. On the rocks and also on the sandy bottom between the rocks, lime-incrustating algae are conspicuous, and *Millepora alcicornis* and *Acropora palmata* are also present.

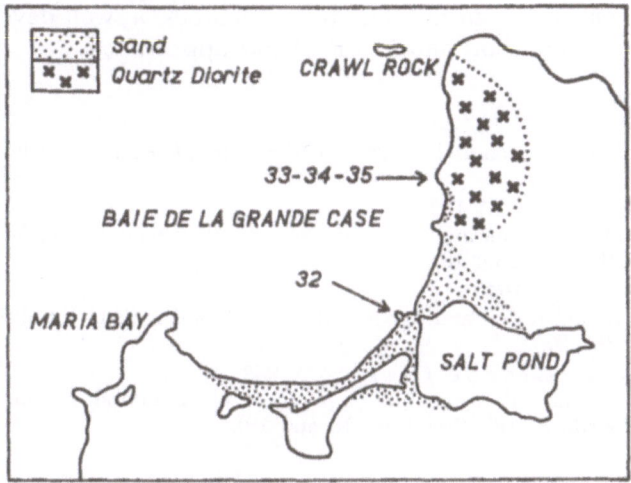

Fig. 18. St. Martin: situation of the sampling spots in Baie de la Case.

Above water level, up to 25 cm, the rocks show small white crusts, probably dead Lithothamnia. A number of algae is rather abundant: *Sargassum polyceratium, Turbinaria turbinata, Laurencia papillosa* and *Cladophoropsis membranacea. Padina sanctae-crucis* is not numerous; of *Chaetomorpha media* only a few specimens have been found. *Lophosiphonia* spec. forms large masses in the most exposed places. Near the water level and immediately below it, several plants of *Dictyota dentata* and *Dilophus guineensis* have been collected. Together with *Cladophoropsis membranacea* were observed: *Laurencia obtusa, L. microcladia, Valonia ocellata, Jania adhaerens,* and *Struvea anastomosans.*

2. Baie de la Grande Case, sublittoral region

In the sublittoral region of Baie de la Grande Case (St. M. 34) numerous algae are present. In the sand, partly covering the rocks, the following species are found: *Caulerpa sertularioides, Halimeda discoidea, H. incrassata, H. opuntia, Penicillus lamourouxii, P. pyriformis, Rhipocephalus phoenix* and *Udotea flabellum.* Attached to the rocks are *Cladophoropsis membranacea, Neomeris annulata, N. mucosa, Valonia ventricosa, Dictyota bartayresii, Dilophus alternans, D. guineensis, Padina sanctae-crucis, Sargassum platycarpum, Amphiroa fragilissima, Centroceras clavulatum, Ceramium nitens, Coelothrix irregularis, Champia parvula, Dasya pedicellata, Digenea simplex, Jania adhaerens, Galaxaura rugosa, G. subverticillata, Laurencia gemmifera,*

L. obtusa, L. papillosa, Liagora ceranoides, L. farinosa, L. pinnata, L. valida and *Polysiphonia ferulacea.* A number of these species has been also mentioned for the upper part of the sublittoral region (St. M. 33).

B. 6. ALGAL VEGETATIONS ON BEACH ROCK (*St. Martin*)

Beachrock is found in many places of St. Martin, mostly in the Low Lands. It is formed in the intertidal zone; its upper limit coincides with the maximum height of normal waves. Behind the plates of beachrock there is often a narrow channel with a sandy bottom; at the seaside there is an abrupt drop. The water of the gully behind the beachrock is constantly renewed through channels in the rock, making this gully a good habitat for many organisms. On the beachrock a well-developed algal vegetation is found, short and of a moss-like appearance.

L e g e n d a :

1. C u p e c o y B a y, 16.V.1958.
Heavy swell; littoral region and upper part of sublittoral region. — Collecting number: St. Martin 22.

2. B u r g e u x B a y, 7.V.1958.
Heavy swell; littoral region and upper part of sublittoral region. — Collecting number: St. Martin 10 (Pl. VIa; Fig. 16).

3. C o l e B a y, 8.V.1958.
Heavy swell; littoral region and upper part of sublittoral region. — Collecting number: St. Martin 11 (Pl. Va).

4. B a i e d e l a G r a n d e C a s e, 24.V.1958.
Slightly exposed; littoral region and upper part of sublittoral region. — Collecting number: St. Martin 32 (Pl. IXa; Figs. 18 and 19).

1. C u p e c o y B a y

In Cupecoy Bay (St. M. 22) the slabs of beachrock before the sandy beach are about 1 m high, but the heavy swell in most places washes high over the rocks. Gastropods are hardly present, only a few specimens of *Tectarius* and *Littorina* are found on more inward rocks. Chitons are absent. Abundant is *Chondria tenuissima.*

A short, moss-like growth of algae consists of: *Polysiphonia ferulacea, Lophosiphonia cristata, Jania adhaerens, Centroceras clavulatum, Cladophora fuliginosa, Cladophoropsis membranacea, Dilophus guineensis, Padina sanctae-crucis* and *Spyridia aculeata.* The development of conspicuously long trichoblasts in part of the material of *Lophosiphonia* and *Polysiphonia* probably depends on the degree of exposure to wave action in this habitat. *Dilophus guineensis, Spyridia aculeata* and *Avrainvillea rawsonii* grow in more or less protected places, in rock fissures, etc. Furthermore, *Caulerpa ambigua, C. sertularioides, Halimeda opuntia, Cladophora uncinata* (between *Polysiphonia*), *Ectocarpus confervoides, Chondria curvilineata, Dichothrix fucicola* and *Hydrocoleum lyngbyaceum* have been collected.

2. B u r g e u x B a y

In Burgeux Bay (St. M. 10), a number of flat slabs of beachrock is formed, with a steep side at the seaside. Behind these slabs small pools are present, 10—20 cm deep, with a sandy bottom. In these pools a.o. *Rhipocephalus phoenix, Halimeda discoidea* and *H. opuntia* are found. As the observations on St. M. 10 for the greater part concern rocks of the Low Lands formation, an accurate description cannot be given of the algal vegetation on beachrock (cf. IV B 2).

3. Cole Bay

Along the beach of Cole Bay (St. M. 11) a number of slabs of beachrock are present, containing hard, white rock fragments (probably derived from the Low Lands formation). Algae are scarce and moss-like. The larger algae grow in crevices of the beachrock and at the seaward side on vertical surfaces. Chitons are absent; *Nerita* is present. At the seaside of the beachrock we find i.a.: *Dilophus alternans*, *D. guineensis*, *Padina sanctae-crucis*, *Sargassum* spec., *Bryothamnion triquetrum* and *Gracilaria debilis*; in fissures of the rock *Cladophora fuliginosa* and *Dictyosphaeria cavernosa*. The moss-like vegetation on the upper side of the beachrock comprises: *Cladophora fuliginosa*, *Cladophoropsis membranacea*, *Padina sanctae-crucis*, *Lophosiphonia cristata*, *Polysiphonia ferulacea*, and other species.

4. Baie de la Grande Case

In Baie de la Grande Case (St. M. 32) the beachrock is less exposed than in the foregoing places (Pl. IXa). Unfortunately our material got lost and the field notes are the only references.

At the seaward side of the beachrock slabs, which are found over a distance of ca. 20 m, a nearly completely closed algal cover is present (Fig. 19).

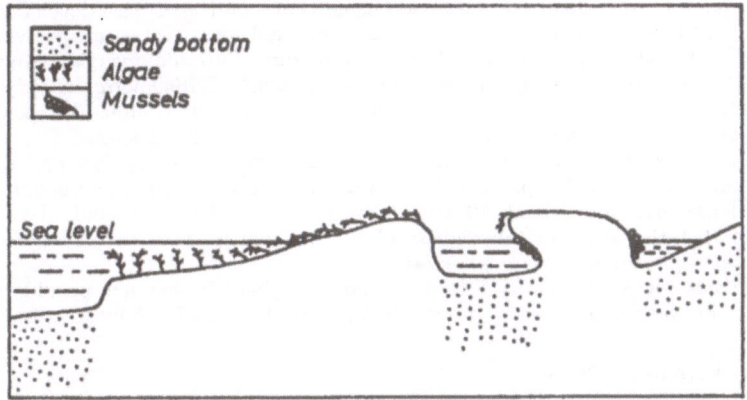

Fig. 19. St. Martin, Baie de la Grande Case: algal zonation on beachrock.

Several species of *Sargassum* are abundant from 30 cm below water level to 10 cm above. *Polysiphonia ferulacea* is abundant, especially in the splash zone above the water level, covering several square meters. *Dictyota* spec. and *Padina* cf. *sanctae-crucis* are rather plentiful, about in the same places as *Sargassum*.

Parallel to the coast a gully is formed in the beachrock, 30–50 cm wide. Against its sides *Cladophoropsis membranacea* is rather abundant; on its sandy bottom *Halimeda opuntia*, *Caulerpa sertularioides*, *Neomeris annulata* (rather common at 10 cm) and *Avrainvillea rawsoni* (rather a lot at water level).

B. 7. Algal vegetations on andesitic rock (St. *Eustatius* and *Saba*)

On a two day visit to S t. *E u s t a t i u s* only a part of the coast could be investigated.

Legenda:

1. B o e k a n i e r s B a y, 20.V.1958.
Rather heavy swell; littoral region and upper part of the sublittoral region. — Collecting number: St. Eustatius 1.

2. B a y b e t w e e n B o e k a n i e r s B a y a n d C o r r e C o r r e B a y, 20.V.1958.
Rather heavy swell; littoral region and upper part of sublittoral region. — Collecting
number: St. Eustatius 2.

3. C o r r e C o r r e B a y, 20.V.1958.
Quiet water, behind coral reef; littoral region and upper part of sublittoral region. —
Collecting number: St. Eustatius 3.

4. C o n c o r d i a B a y, 20.V.1958.
Rocks in front of sandy beach; rather heavy swell; littoral region and upper part
of sublitoral region. — Collecting number: St. Eustatius 4.

5. B a c k - o f f B a y, 21.V.1958.
Rocks along coast near Sugarloaf. Slight wave action; littoral region and upper part
of sublittoral region. — Collecting number: St. Eustatius 5 (Pl. X).

6. B a c k - o f f B a y, 21.V.1958.
Slight wave action; sublittoral region. — Collecting number: St. Eustatius 6.

The main part of the coast of St. Eustatius consists of large boulders; in several
places, mainly in bays, there is a sandy beach, usually with many pebbles. The
boulders lie in front of a steep cliff of layered tuffs; they are rounded and have a
rough surface; their colour varies from dark grey to light grey and brown.

In several places along the coast a magnificent coral reef has developed (Corre
Corre Bay, St. Eust. 3). Behind the reef, in quiet and clear water, many living corals
can be found, i.a. *Acropora palmata, A. cervicornis, Eusmilia fastigiata* and *Porites
porites*. In such places many urchins are also present: *Echinometra* in very shallow
water, somewhat deeper *Diadema* and *Lytechinus*. Numerous Chitons are seen; several
species of *Nerita* are abundant; *Littorina* and *Tectarius* are numerous.

The water level, on the 20th and 21st of May 1958, was very low and there was
hardly any wind. In Schildpadden Bay (Turtle Bay) elkhorn coral, *Acropora palmata*,
over a large area rose about 10 cm above the water level. It could be distinctly
observed that the algae in the highest places had partly died, including *Turbinaria,
Sargassum* spp. and *Laurencia papillosa*.

The notes for St. Eustatius 1 to 4 are not complete. It was not possible to make
notes at the spot and the trip proved to be very long and exhausting.

1. B o e k a n i e r s B a y

Near Boekaniers Bay (St. Eust. 1) the coast is formed by large rounded rocks with
a dense algal vegetation to high above sea-level. Above the algal zone the colour
of the rocks is somewhat lighter. At the upper margin of this light-coloured zone
Chitons are abundant. Still higher the rocks have their normal colour. *Turbinaria,
Sargassum* spec., *Laurencia microcladia, L. papillosa, Grateloupia cuneifolia, Polysiphonia ferulacea, Dictyota ciliolata* and *Dilophus alternans* are plentiful. *Grateloupia
cuneifolia* grows in places where the waves splash the rock, and *Spyridia aculeata* is
also rather common in these places. *Chaetomorpha media* is abundant about the low
water mark. On wet rocks large masses are formed of *Lophosiphonia* spec. *Bostrychia*,
usually found in such places, was not observed. *Pocockiella variegata* and crustlike
Lithothamnia are abundant, and several specimens were noticed of *Centroceras clavulatum, Hypnea musciformis* and *Cladophoropsis membranacea*.

2. B a y b e t w e e n B o e k a n i e r s B a y a n d C o r r e C o r r e B a y

From Boekaniers Bay to Corre Corre Bay (St. Eust. 2) the habitat factors change
only slightly.

In small pools between the rocks, somewhat less exposed to wave action, have
been found: *Amphiroa fragilissima, Dictyota dentata, D. jamaicensis, Chaetomorpha
media* and *Valonia ventricosa. Sargassum platycarpum, S.* cf. *vulgare* and *Stypopodium
zonale* were washed ashore.

3. Corre Corre Bay

Behind the beautiful coral reef of Corre Corre Bay (St. Eust. 3) there is a shallow, sheltered bay of a maximum width of 75 m. The water is shallow and very clear, and contains many corals. Because wave action is completely broken by the coral reef, the algal flora resembles that of a pool or lagoon. Only a few remarkable species will be mentioned. *Turbinaria* is abundant; *Caulerpa racemosa* is rather numerous in the upper part of the sublittoral region, just like *Galaxaura cylindrica* and *G. subverticillata* (about 10 cm deep), and *Liagora valida*. On rocks, washed by the waves, *Ulva fasciata* has been collected; *Gelidiella acerosa* is rather common. The genus *Laurencia* is represented by *L. gemmifera*, *L. microcladia*, *L. obtusa*, *L. intricata* and *L. papillosa*. Several species grow together in a dense vegetation: *Laurencia intricata*, *Jania adhaerens*, *Ceramium* spec., *Centroceras clavulatum*, *Hypnea* spec., *Grateloupia filicina*.

4. Concordia Bay

Concordia Bay (St. Eust. 4) has a long sandy beach in front of a cliff, about 10 m high. Several large boulders are found with a dense algal vegetation, washed by the sand-loaded waves. Much *Sargassum natans* is washed ashore, together with *Dictyopteris justii* and *Ochtodes filiformis*. Between the rocks *Halimeda tuna* is found.

On the rocks, just below the water level several specimens of *Bryothamnion triquetrum* are present, constantly scoured by sand; *Padina* cf. *vickersiae* is not abundant. On the boulders also: *Dictyosphaeria vanbosseae*, *Laurencia* spec., *Digenea simplex*, *Dictyota dentata*, *Dilophus guineensis*. Epiphytes are present, especially on *Digenea simplex*, *Laurencia* spec. and *Dictyopteris justii*. The following species have been observed: *Dipterosiphonia dendritica*, *Jania adhaerens*, *Hypnea musciformis*, *Champia parvula*, *Polysiphonia ferulacea*, *Ceramium byssoideum*, *Lophosiphonia cristata*, *Spermothamnion turneri*, *S. investiens* and *Diplochaete solitaria*.

5. Back-off Bay

In Back-off Bay, at the base of Sugarloaf and White Wall (St. Eust. 5 and 6; Pl. X), the coast is formed by boulders over about 100 m. White Wall is partly adjacent to the sea, but this part proved to be inaccessible. The bottom slopes down rather steeply and at about 20 m from the coast the rocks rest in a sandy bottom at a depth of about 5 m. Numerous black sea cucumbers appeared to be almost the only organisms to 10 cm below sea-level (extreme low water).

Sargassum spec. and *Turbinaria* are not abundant. About the water level a great number of *Dictyota ciliolata*, *D. dentata*, *Dilophus guineensis*, *Chaetomorpha media* and *Enteromorpha* spec. are present. This algal vegetation is partly dense and contains many intermingled species: *Polysiphonia ferulacea*, *Spyridia aculeata*, *Chaetomorpha media*, *Padina* cf. *vickersiae*, *Laurencia microcladia*, *Pocockiella variegata*, *Hypnea musciformis*, *Grateloupia cuneifolia*, *Wrangelia argus*, *Lophosiphonia cristata*, *Stichothamnion antillarum* and *Laurencia papillosa*. Several of the larger species bear epiphytes, e.g. *Melobesia farinosa*, *Diplochaete solitaria*, *Herposiphonia tenella*, *Ceramium* spec. and *Jania* spec.

6. Back-off Bay, sublittoral region

Though the sublittoral vegetation is not very rich, four species of *Galaxaura*, viz. *G. cylindrica*, *G. oblongata*, *G. rugosa* and *G. subverticillata*, are found together with *Amphiroa hancockii* at a depth of 1—3 m.

Also on **Saba** (Fig. 3) samples could be collected for two days only. During a boattrip round the island a fairly accurate general picture of the coast was obtained.

Legenda:

1. Washgut, 27.V.1958.
Rocks along the coast. Rather heavy swell; littoral and sublittoral region. — Collecting number: Saba 1.

2. Eastern coast of Saba, near The Level, 27.V.1958.
Very heavy swell; sublittoral region. — Collecting number: Saba 2.

3. Flat Point, 27.V.1958.
Shallow rock pool. Heavy swell; littoral region and upper part of sublittoral region. — Collecting number: Saba 3 (Fig. 20).

4. Northern coast, near Old Sulphur Mines, 27.V.1958.
Large rock, about 10 m off shore. Heavy swell; sublittoral region. — Collecting number: Saba 4.

5. Fort Bay, 28.V.1958.
Near customs-office towards the west. Rocks along the shore-line; in front of it a sandy bottom. Slight wave action; sublittoral region. — Collecting number: Saba 5.

6. West of Fort Bay, 28.V.1958.
Rocks along the shore. Slight wave action; littoral region and upper part of sublittoral region. — Collecting number: Saba 6 (Pl. IXb; Fig. 21).

7. Southwestern coast, at the base of Parish Hill, 28.V.1958.
Several large boulders. Slight swell; littoral region and upper part of the sublittoral region. — Collecting number: Saba 7.

8. Southwestern coast, at the base of Parish Hill, 28.V.1958.
See preceding number; sublittoral region. — Collecting number: Saba 8.

The coast of Saba is rather monotonous. The bottom of the sea slopes down steeply and is largely formed by large boulders of andesitic origin. The neighbourhood of Fort Bay in front of the coast has a clear sand-bottom.

From Fort Bay eastward the coast consists of boulders and is rather uniform. The boulders in the splash zone are rather rich in algae; the lower parts only have a moss-like vegetation.

East of The Level a steep cliff is formed. This type of coast may be seen as far as Spring Bay. Old Booby Hill rises also vertically from the sea. On this rocky coast, normally heavily exposed to the waves, a distinct zonation can be seen. From above downwards the following zones may be distinguished: original rock — dark-coloured zone — light coloured zone — algal zone. The upper limit of the dark-coloured zone is also the upper limit of the splash zone. Chitons clearly indicate the upper limit of the light-coloured zone; in some places they reach a little higher.

At Spring Bay the steep cliff passes into a boulder beach; near Flat Point the cliff again borders the sea. Flat Point, however, is not very high and in places the rock slopes down at an angle of 30°. Consequently, the waves run high on the coast and a broad zone of seaweeds is formed in the littoral region and the upper part of the sublittoral region.

Further to the west the northern coast is steep again. Near Diamond Rock and Torrens Point, the rocks rise vertically from the sea to a considerable height; deep guts are carving the slopes of The Mountain. The shore-line consists of boulders of various sizes, often completely barren and in constant motion.

1. Washgut

Near Washgut (Saba 1) the bottom slopes steeply. On large boulders *Amphiroa fragilissima* and a bluegreen alga are found. For the rest there are almost only animals, mainly corals, and gorgonids.

Boulders on the shore-line have a well-developed algal vegetation: *Laurencia papillosa*, *Polysiphonia ferulacea* and *Sargassum* are numerous, and several specimens were found of *Chaetomorpha media*, *Turbinaria*, *Grateloupia cuneifolia*, *Hypnea*

62

musciformis, *Laurencia microcladia* and *L. intricata*. Conspicuous is also a small species of *Laurencia* which in many places forms flat growths on the rock. Its thallus is only about 1 cm high and only branched at the tips. Possibly this may be an ecological form of *L. papillosa*.

2. East coast of Saba, between Washgut and Curve Gut

Between Washgut and Curve Gut, near The Level (Saba 2), the coast is very steep. The waves beat high against the rocky shore, which proved to be completely inaccessible. The sublittoral region is very poor in algae. Only *Grateloupia cuneifolia* and a few bluegreen algae have been recognized. For the rest mainly animal organisms are present on the greyish-red andesitic rocks.

3. Flat Point

Near Flat Point (Saba 3) the coast rises steeply from the sea to about 1 m, then a flat terrace of 5—10 m width is formed. At the front side there is a small increment, and behind this a sheet of water, 10—20 cm deep, which is constantly renewed by the waves. At the landside of the terrace a gully of about 1½ m depth is present, in which corals are growing, mainly *Millepora alcicornis*, but no algae (Fig. 20). The front side of the terrace shows a zone of barnacles; behind the threshold grow numerous plants of *Sargassum platycarpum*, *S. polyceratium* and *Turbinaria*.

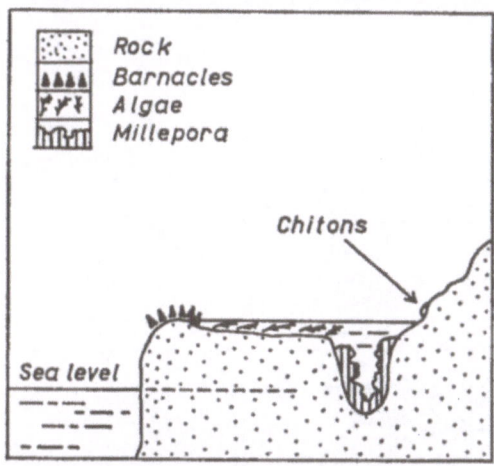

Fig. 20. SABA, Flat Point: algal vegetation, barnacles and stinging coral in a rock pool.

The vegetation is formed by many intermingled species: *Laurencia papillosa*, *L. gemmifera*, *L.* cf. *obtusa*, *Laurencia* spec. (same as found in Washgut), *Polysiphonia ferulacea*, *Padina gymnospora*, *Dilophus guineensis*, *Pocockiella variegata*, *Centroceras clavulatum* and several bluegreen algae, i.a. *Symploca hydnoides*, forming thick cushions. On larger algae several epiphytes can be observed: *Ceramium byssoideum*, *Dasya* spec., *Herposiphonia* spec. and *Lyngbya majuscula*.

4. Northern coast of Saba, near Old Sulphur Mines

The northern coast, near the Old Sulphur Mines (Saba 4) is inaccessible. The shore-line and also the seabottom is formed by large boulders. On one of these, about 10 m in front of the shore, from the surface to a depth of several meters, two species of algae are abundant: *Dictyopteris delicatula* and *Dictyota jamaicensis*.

Just as on the eastern coast of Saba, a zonation is visible also along the northern coast over several hundreds of meters.

5. Fort Bay

On boulders immediately in front of the coast (Saba 5) a short, moss-like vegetation is developed with remarkably much *Amphiroa fragilissima*. At a depth of 1½–3 m several specimens of *Galaxaura oblongata*, *G. rugosa*, and *G. subverticillata* are found, with a.o. *Champia parvula* and *Dictyopteris delicatula* as epiphytes. The moss-like vegetation contains also *Colpomenia sinuosa*, *Padina vickersiae*, *Hypnea* spec., *Jania* spec., *Grateloupia* spec., *Laurencia papillosa* and *Wrangelia argus*.

6. West of Fort Bay

Several hundreds of meters west of Fort Bay (Saba 6) the coast is also formed by large boulders. Only a few larger algae are found in the sublittoral region; the zone regularly washed by the waves, however, presents a rich fauna and flora. Chitons are very numerous, most of them at the upperside of a light-coloured zone in the rocks, in some places a little higher (Fig. 21). *Nerita tessellata* is abundant, from the water level to the light-coloured zone. Between the rocks many Echinometras are

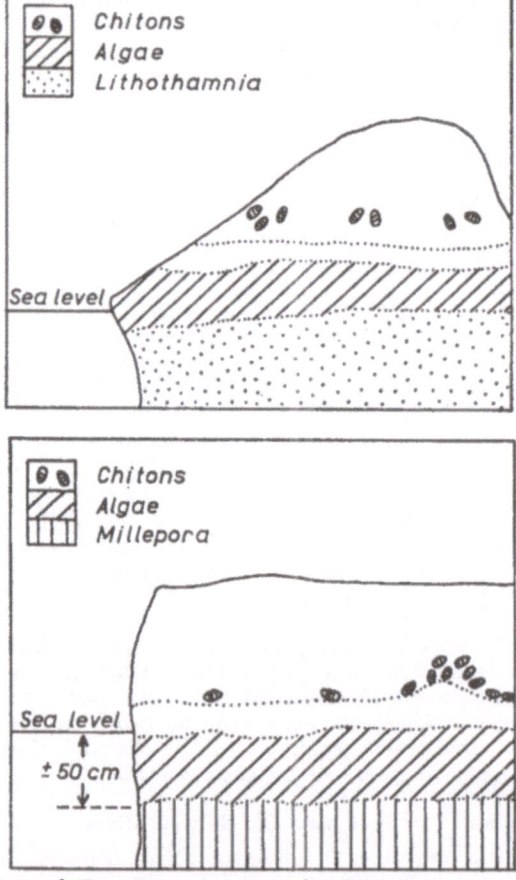

Fig. 21. Saba, west of Fort Bay: zonation of organisms on two andesitic boulders, illustrating the fact that Lithothamnia and corals exclude each other. The differences in the upper limit of the organisms in relation to the sea-level are caused by different exposure to wave action.

present. The surface of the rocks at this level is mostly covered with crustlike Lithothamnia. *Pocockiella* is less abundant. Lithothamnia and *Millepora* exclude each other (Fig. 21). High on the rocks, seldom or never wetted by the waves, *Littorina* and *Tectarius* are abundant.

In the algal vegetation, partly again short and moss-like, *Sargassum* cf. *vulgare*, *Turbinaria* and *Polysiphonia ferulacea* are numerous. *Laurencia* spec. forms large patches on the sides of the rock. *Laurencia microcladia*, *Dictyota* spec. and *Wrangelia argus* can be found also; the latter immediately below the water level. On strongly exposed places *Sargassum vulgare*, *Turbinaria*, *Laurencia papillosa*, *L. microcladia*, *L. intricata*, *Grateloupia cuneifolia* and *Centroceras clavulatum* may be observed.

7. Southwestern point of Saba, near Parish Hill

The southwestern point of Saba, near Parish Hill (Saba 7) shows many small boulders. Only a few larger boulders, somewhat off-shore and projecting a little above sea-level, have the same species as in other places.

8. Southwestern point of Saba, sublittoral region

In the sublittoral region (Saba 8) on rocks near the shore only a short moss-like algal vegetation has developed. Noteworthy are *Amphiroa fragilissima*, *Gelidiella acerosa*, *Pterocladia pinnata*, *Wrangelia argus* and several bluegreen algae. Rocks, further off-shore, have more algae; on several of them *Liagora decussata* is abundant.

C. ALGAL VEGETATIONS OF SANDY BEACHES AND LAGOONS

The sea around St. Martin is generally shallow. The island, together with Anguilla, Fourche, and St. Barts, is situated on a shallow bank; during the ice-ages all these islands together formed one large island, known as Great St. Martin (Fig. 1). Shallow sandy bottoms are found in Long Bay, Simson Bay, Little Bay, Great Bay, Mouth Piece Bay (Anse de l'Embouchure), Baie de la Grande Case and Marigot Bay (Fig. 2). In most cases they have extensive vegetations of seagrasses. Large stretches of the shore of St. Martin consist of finegrained sandy material, particularly in the bays on the southern and northwestern coast. Beaches consisting of coral debris, common on Curaçao, Aruba and Bonaire, do not occur on St. Martin.

The large lagoons and some smaller ones are generally separated from the sea by long, slightly curved, sandy bars, e.g. near Simson Bay Lagoon, Great Salt Pond, Salt Pond near Grande Case, Fish Pond. Only a few of the lagoons of St. Martin are widely connected to the sea, viz. Fish Pond and Oyster Pond. Formerly, Simson Bay Lagoon at its south end (Fig. 16, near arrow for 1129) also had an opening to the sea. It was closed, however, between 1950 and 1955. As a result of this many of the mangroves bordering the lagoon were dying in 1958.

The seagrass and algal vegetation around St. Martin was studied in several places, and also in a number of lagoons, viz. Fish Pond, Oyster Pond and Simson Bay Lagoon (inclusive Mullet Pond and Flamingo Pond), samples were taken. Sandy bottom habitats around St. Eustatius and Saba are scarce and, therefore, were not studied. Saba and St. Eustatius have no lagoons.

C. 1. SANDY BEACHES (*St. Martin*)

Many places along the coast of St. Martin have sandy beaches, especially along the southern and nothwestern coast (Pls. Ib, II, IIIb, V. VI, VIIIa, IXa).

The degree of exposure to wave action may differ considerably. Along the eastern coast of St. Martin, and also in the bays of the southwestern part of the island, a heavy swell causes the beach often to be steeply eroded to 1 or 1½ m. At a higher level a land vegetation immediately is found: STOFFERS' (1956) "herbaceous strand community" and "sand scrub community", occurring at Simson Bay, Mullet Pond Bay, Guana Bay and Oyster Pond.

Only animals are found in the littoral region and at higher levels; algae are only present in the sublittoral region, where the bottom is more stable. In gullies behind the plates of beachrock formed on many sandy beaches, certain algae may be abundant. The water of these gullies is regularly renewed by the waves (Fig. 19).

Legenda:

1. S i m s o n B a y, eastern part, 6.V.1958.
Sandy bottom; sublittoral region. — Collecting number: St. Martin 9 (Fig. 16).

2. G r e a t B a y, near Little Bay Hotel, 22.V.1958.
Sandy bottom; sublittoral region. — Collecting number: St. Martin 24 (Pl. IIIb; Fig. 14).

3. M o u t h P i e c e B a y (Anse de L'Embouchure), 23.V.1958.
Sandy bottom; sublittoral region. — Collecting number: St. Martin 28 (Fig. 15).

4. B a i e d e l a G r a n d e C a s e, 26.V.1958.
Sandy bottom; sublittoral region. — Collecting number: St. Martin 35 (Fig. 18).

5. L i t t l e B a y, eastern part, off Fort Amsterdam, 30.IV.1958.
Sandy bottom with boulders of different origin, more outwards a clean sandy bottom; sublittoral region. — Collecting number: St. Martin 4 (Pl. IIIb; Fig. 14).

6. S i m s o n B a y, eastern part, 6.V.1958.
Sandy bottom with boulders, uppermost part of the sublittoral region. — Collecting number: St. Martin 8 (Pl. Vb; Fig. 16).

7. G r e a t B a y, near the landing-stage at Point Blanche, 14.V.1958.
Sandy bottom, partly with boulders; sublittoral region. — Collecting number: St. Martin 21 (Fig. 12).

8. B a i e d e l a G r a n d e C a s e, 26.V.1958.
Sandy bottom with dioritic rocks; uppermost part of sublittoral region. — Collecting number: St. Martin 34 (Fig. 18).

The stations may be divided into two groups. The first four stations are in places with a clean sandy bottom of a certain vastness. The second four are in places where boulders of different sizes rest in a sandy bottom. Partly these boulders are covered by a thin layer of sand. Between the boulders the water is mostly in heavy motion and carries much sand. The algae between the boulders mostly grow also at greater depth.

1. S i m s o n B a y

The eastern part of Simson Bay (St. M. 9) has a clean sandy bottom. The place of sampling was 50 x 50 m, and 50—100 cm deep. The seagrasses grow in patches, mostly rising above the surrounding sandy bottom, occasionally up to 30 cm. The bottom is barren between and the sand is constantly moving. *Syringodium* is abundant;

Thalassia appears to be less common.

The following algae are present among the seagrasses: *Caulerpa sertularioides, Halimeda discoidea, H. incrassata* var. *monilis* f. *cylindrica, H. incrassata* var. *monilis* f. *robusta, H. incrassata* var. *simulans, Penicillus capitatus, P. lamourouxii* and *Udotea flabellum.* Also *Padina sanctae-crucis* and *Laurencia obtusa* are present. The following epiphytes are found on the seagrasses, especially *Thalassia,* and also on a number of larger algae: *Entocladia* spec., *Ulvella lens, Centroceras clavulatum, Chondria curvilineata, Crouania* spec., *Hypnea* cf. *cervicornis, Herposiphonia* spec., *Melobesia farinosa, Polysiphonia binneyi* and *P.* cf. *gorgoniae.*

Some boulders in the sandy bottom carry several specimens of *Neomeris annulata,* and *Dictyosphaeria cavernosa* is rather abundant.

The vegetation of Simson Bay is of a great extent. Most of the detached leaves of the seagrasses is washed ashore in the western part of the bay, where a wall of *Syringodium* has been deposited, at least ½ m high and 1–2 m wide (St. Martin 10a).

2. Great Bay

The western shore of Great Bay (St. M. 24) has a heavy swell. During our observation, however, the weather was quiet and part of the bay, to about 50 m off-shore, could be investigated. At 10 m off-shore the water has a depth of about 50 cm and the sandy bottom is visible between the doleritic rocks (See chapter IV B 4). Still more outward a bottom of clean sand is found. About 50 m off-shore the water is about 2 m deep. In deeper water the vegetation, mostly *Thalassia,* diminishes and also other organisms become less numerous. In a few places, e.g. near the entrance gate of Little Bay Hotel, the vegetation consists exclusively of *Syringodium.*

The patches of *Thalassia* contain a lot of *Halimeda opuntia* and *H. discoidea.* Particularly *Halimeda opuntia* covers large areas of the bottom. *Halimeda discoidea* grows in comparable places, but probably in deeper water than *H. opuntia.* Only a few plants of *Halimeda incrassata* var. *simulans* have been found. *Penicillus capitatus* is not abundant. *Penicillus dumetosus,* however, is numerous in the *Thalassia* and *Halimeda* vegetation at a depth of 1–2 m. *Udotea conglutinata* and *U. flabellum* are also present. *Caulerpa racemosa* is abundant; robust specimens are found growing over and between *Halimeda* at depths varying from 0 to 200 cm. Several large plants of *Caulerpa prolifera* have been found, mostly in places where the growth of *Thalassia* is less dense. *Caulerpa sertularioides* and *C. taxifolia* are found in places, comparable with those with *Caulerpa racemosa.* The brown algae appear to be poorly represented. *Dictyota divaricata* is abundant between *Thalassia* and *Halimeda.*

3. Mouth Piece Bay

The bottom of the bay is sandy and carries an extensive growth of algae and seagrasses. Part of it will be described in greater detail (St. M. 28).

In the first 15 m from the coast, banks closely covered with *Thalassia* and *Syringodium* are formed, several square meters in size and rising to about 50 cm above the surrounding sandy bottom. More seaward the bottom is flat with the same dense vegetation. At a depth of about 2 m, the banks rise to about 1 m below the water level. Algae are scarce in this dense vegetation of seagrass (mainly *Halimeda*-species and *Udotea flabellum*). *Lytechinus variegatus* is numerous; *Oreaster reticulatus* is present, and several not full-grown specimens of *Strombus gigas* are present among *Thalassia.* This species is collected for consumption, as is clear from many empty shells on the beach.

The algal vegetation in this bay is not luxuriantly developed. *Penicillus capitatus,* however, is abundant, mostly in the seagrass vegetation beginning at about 15 m off-shore, and on the banks. *Halimeda incrassata* var. *monilis* f. *robusta* is abundant in flat parts; *Halimeda opuntia* mainly on the banks. *Udotea flabellum* grows in both places, but is not numerous. Large masses of *Valonia aegagropila,* with a diameter of 5–10 cm are present on the banks and in the gullies between them, generally together with other *Valonia*-species. *Ernodesmis verticillata* is rather abundant. *Caulerpa sertu-*

larioides and *C. cupressoides* are present in several places on the bottom between the grasses. A few specimens of *Caulerpa cupressoides* have been found at about 2 m.

Thalassia and *Syringodium* and also the larger seaweeds carry a number of epiphytes: *Amphiroa fragilissima, Jania adhaerens, Melobesia farinosa* (the latter is also abundant on many algae).

In some places a few boulders are found, rising somewhat above sealevel. They carry *Laurencia papillosa, Sargassum* spec. and *Turbinaria. Chondria tenuissima* is found at about 30 cm.

4. Baie de la Grande Case

Also in this locality (St. M. 35) the bottom is sandy; the depth of the sea appears to be about 2 m over hundreds of meters. From the coast many dark-coloured patches of seagrasses may be observed. *Syringodium* is more plentiful than *Thalassia*. Even in places without seagrasses, algae are present here, particularly *Padina sanctae-crucis, Dilophus guineensis, Dictyota cervicornis* and *D. indica*. More regularly distributed, also among the seagrasses, are *Halimeda incrassata* var. *monilis* f. *robusta, Penicillus capitatus* and *Udotea flabellum*, together with many red algae. Epiphytically on the seagrasses and large algae have been observed: *Ceramium floridanum, Chondria* cf. *atropurpurea, Dasya rigidula, Lophosiphonia* spec., and *Polysiphonia ferulacea*.

5. Little Bay, eastern part, off Fort Amsterdam

In the eastern part of Little Bay (St. M. 4) the boulders rest in a sandy bottom; many of them are often covered with sand. *Caulerpa sertularioides* is abundant in the sandy bottom and also in the thin layer of sand covering the boulders at a depth of 20–100 cm. *Penicillus capitatus* and *P. lamourouxii* are numerous in similar places, 20–50 cm deep, but not well-developed.

Several species, normally requiring a hard substratum for attachment, appear to grow here on rocks covered with a rather thick layer of sand. Only one specimen of *Acetabularia calyculus* was found at about 30 cm: *Neomeris annulata* is abundant near and just below the low water mark. When deeper than 1 meter, the bottom of Little Bay consists of sand, without any vegetation.

6. Simson Bay, eastern part

The eastern part of Simson Bay (St. M. 8) has about the same conditions as Little Bay. The algal vegetation on the boulders in Simson Bay has been discussed in chapter IV B 2.

In shallow sandy parts between the boulders *Caulerpa cupressoides, C. racemosa, C. sertularioides, Penicillus capitatus, P. lamourouxii, Halimeda incrassata, H. opuntia* and *Rhipocephalus phoenix* have been collected. These plants appear to be better developed in Simson Bay than in Little Bay.

7. Great Bay, eastern part, near Point Blanche

The algal vegetation of Great Bay, near Point Blanche (St. M. 21) has been described in chapter IV B 3.

8. Baie de la Grande Case

The algal vegetation of Baie de la Grande Case (St. M. 34) has been described in chapter IV B 5.

C. 2. LAGOONS *(St. Martin)*

Only a few of the lagoons of St. Martin are widely connected to the sea, viz. Fish Pond and Oyster Pond. The entrance of Simson Bay Lagoon was barred a few years before the date of collecting, and the flora and fauna of this large lagoon proved to be already distinctly improverished. The lagoons are bordered by "mangrove woodland", in which locally a distinction may be made between a *Rhizophora* consocies and an *Avicennia* consocies, strongly influenced by man (STOFFERS, 1956).

Legenda:

1. O y s t e r P o n d, 12.V.1958.
Mangroves along the southern bank, sandy bottom; littoral region and upper part of the sublittoral region. — Collecting number: St. Martin 15 (Fig. 15).

2. F i s h P o n d, 23.V.1958.
Mangrove vegetation near the entrance of the lagoon in Mouth Piece Bay (Anse de l'Embouchure); littoral and sublittoral region. — Collecting number: St. Martin 26 (Fig. 15).

3. S i m s o n B a y L a g o o n, Flamingo Pond, 24.V.1958.
Shore vegetation with *Rhizophora* and *Avicennia*; silty bottom, covering rocks of the Low Lands formation. — Collecting number: St. Martin 29 (Fig. 16).

4. S i m s o n B a y L a g o o n, Mullet Pond, 9.V.1958.
Collecting number: St. Martin 29a (Fig. 16).

5. S i m s o n B a y L a g o o n, southwestern part, near Mary Point, 10.V.1958.
Shore vegetation with much *Batis*, in front of it *Avicennia* in silty bottom. — Collecting number: St. Martin 29b (Fig. 16).

6. S i m s o n B a y L a g o o n, 24.V.1958.
Rhizophora vegetation, several tens of meters east of buildings of Juliana Airport, together with several *Avicennia* on a sandy bottom. — Collecting number: St. Martin 30 (Fig. 16).

7. S i m s o n B a y L a g o o n, 24.V.1958.
Sandy bottom near former entrance, with *Salicornia*, *Sesuvium* and *Conocarpus*. Between the banks narrow creeks, 10—20 cm deep. — Collecting number: St. Martin 31 (Fig. 16).

1. Oyster Pond

The algal vegetation of Oyster Pond has partly been described already in the discussion of the vegetation on coral limestone in chapter IV B 1.

Along the southern bank the shore is rather varied. In the southernmost part, where our sample was taken, first a sandy beach is found, with boulders of the Point Blanche formation. Then there follows a small patch with several *Rhizophora* and *Avicennia* in a rather silty bottom. Further to the northeast the shore changes into coral limestone, with a sandy bottom in front of it.

On the *Rhizophora* roots *Bostrychia binderi* and *Polysiphonia howei* are well-developed, just as they are under the overhanging rocks of coral limestone; *Bostrychia* grows at a slightly higher level than *Polysiphonia*. The latter grows about 10 cm above the water level. Near the water level, on the roots of *Rhizophora* and also on rocks, many small barnacles are found. On the roots, among the algae mentioned already, *Ulva lactuca* is abundant, and *Caloglossa leprieurii* and *Murrayella periclados* also occur. *Acanthophora spicifera* is to be found in similar places and also on boulders in shallow water.

In the most inward corner of Oyster Pond, where *Rhizophora* and *Conocarpus* are present, a gully discharges into the lagoon. Here sedimentation is such that algal growth is probably completely absent.

The whole southern shore of the lagoon is formed by sand with many boulders of the Point Blanche formation. In the shallow water a lot of *Thalassia* and *Syringodium* is growing. On rocks *Laurencia papillosa* is common. *Halimeda opuntia* and *Amphiroa fragilissima* are washed ashore.

69

The greater part of Oyster Pond is very shallow; to about 10 m off-shore the depth is only 30—50 cm. In the seagrass vegetation only *Penicillus capitatus* is found. In the silty bottom near the mangroves and in front of the sandy shore in the south-eastern part of the lagoon large specimens of *Padina gymnospora* are present. *Acanthophora spicifera* appears to be abundant on rocks and on the roots of mangroves.

In the sandy bottom in front of the coral limestone a seagrass vegetation has developed. Next to *Penicillus capitatus*, also *Caulerpa sertularioides*, *Halimeda opuntia* and *Amphiroa fragilissima* are found. Over a large area, at a depth of about 40 cm, many loose specimens of *Lithothamnion* are observed. A number of boulders resting in the bottom is completely covered by a pink-coloured, lime-incrusting alga.

2. Outlet of Fish Pond in Mouth Piece Bay

The connection of Fish Pond and Mouth Piece Bay (St. M. 26) is about 15 m wide, more inwardly even as narrow as 10 m; its depth in many places is about 1 m. A small islet is situated in the mouth, thus forming two channels, densely bordered with *Rhizophora* and *Avicennia*. Towards Fish Pond these trees form a dense zone.

The southernmost channel (Fig. 15) comes to a dead end in the mangrove vegetation. The water was stagnant and appeared to be rather cool. The water of the other channel, however, was warmer and ran slowly in the direction of Mouth Piece Bay. Apparently, Fish Pond is rather shallow, causing the water to become very warm. Many small fishes are present, e.g. young barracudas. Near the outlet the water is shallow (dotted in Fig. 15), only 0—25 cm. At 150 m off-shore the depth is only 75 cm, and there a vegetation of *Thalassia* and *Syringodium* is found.

Samples were taken in the channels along the islet in the entrance to the lagoon. No barnacles or oysters were seen on the roots of the mangroves. Everything is covered with mud.

Acanthophora spicifera forms at least 90% of the algal vegetation and grows both on the roots of the mangroves and on rocks along the shore. In the dead-end channel *Halimeda opuntia* and *Udotea flabellum* are rather abundant: *Chaetomorpha brachygona* forms dense masses between other algae and the roots of the mangroves. On the mangroves also *Polysiphonia denudata* and *Centroceras clavulatum* are growing. Epiphytically on *Syringodium*, *Enteromorpha chaetomorphoides* is found, intermingled with *Chondria curvilineata*; on *Chaetomorpha brachygona*: *Diplochaete solitaria*, *Asterocytis ramosa* and *Kylinia crassipes*; and on *Thalassia*: *Melobesia farinosa*.

Along the whole spit of land, separating Fish Pond and Mouth Piece Bay, at the land side, a dense growth of *Rhizophora* and *Avicennia* is developed. The other shore of Fish Pond is devoid of mangroves.

3. Simson Bay Lagoon, Flamingo Pond

The algal vegetation of Simson Bay Lagoon, and also of Flamingo Pond and Mullet Pond (St. M. 29, 30 and 31) appears to be very poor. Owing to the closure of the entrance the lagoon flora and fauna deteriorated and in the eastward part most of the mangrove vegetation died. Before it was closed, Simson Bay Lagoon was a rich fishing area.

The vegetation along the shore of Flamingo Pond consists mainly of *Rhizophora* and *Avicennia* (St. M. 29). Many fiddler crabs are seen along the shore, but the water is poor in fish. The bottom is muddy and almost without plantlife. In a few places small tufts of *Batophora oerstedi* are seen, mostly attached to shells of *Cerithium*.

During our visit only the southern part of Flamingo Pond was studied. The level of the water was rather high, owing to the heavy rains between the 2nd and the 5th of May 1958. A better developed algal vegetation may have been present in deeper parts of the lagoon.

70

4. Simson Bay Lagoon, Mullet Pond

In Mullet Pond (St. M. 29a) *Batophora* is found and probably species of *Enteromorpha*.

5. Simson Bay Lagoon, southwestern part

Along the most southwestern part of Simson Bay Lagoon (St. M. 29b) a rich *Batis* vegetation is found, again partly submerged owing to the exceptional rainfall during the first days of May (1958). More landwards, many Avicennias grow in a muddy bottom. In this part of Simson Bay Lagoon algae were not seen at all.

6. Simson Bay Lagoon, near Juliana Airport

Near the building of Juliana Airport along the shore of Simson Bay Lagoon, well-developed, living Rhizophoras are present with several Avicennias amongst them (St. M. 30). The bottom consists of sand, with numerous fragments of shells and segments of *Halimeda*. In front of the Rhizophoras the lagoon is somewhat deeper, about 50—100 cm, and the bottom particles are smaller. The mangroves have some algal growth. Abundant is *Enteromorpha compressa;* some plants of *Batophora* were seen about 20 cm deep, and in a few places also seagrasses were present.

7. Simson Bay Lagoon, near former outlet

Near the former outlet of Simson Bay Lagoon, on both sides of the bridge, is a wide bank of sand with a rich land vegetation with large patches of *Tournefortia gnaphalodes*. At the lagoon-side of the bridge large sandy flats are present with *Salicornia*, *Sesuvium*, and a few bushes of *Conocarpus*. Small gullies are formed, 10—20 cm deep. Close to the bridge, *Enteromorpha intestinalis* is found. In other parts algae were not observed.

In conclusion, a few remarks on lagoons completely separated from the sea.

Near Little Bay (Fig. 14, north of the arrow for St. M. 6) a small lake of about 100 m in width was formed after the heavy rains in the first days of May. Probably fastgrowing brackish water species can develop well in these temporary lakes.

In Great Salt Pond (Fig. 2) the habitat factors also vary. Algae have not been found.

Near Cole Bay (St. M. 11), at the landside of the cemetery is a lagoon with probably brackish water (due to rainfall). Many fiddler crabs are observed; algae are not present.

Near Maho Bay (St. M. 10) there is also a small pond with numerous fiddler crabs but without algae.

Exactly north of the coast of coral limestone near Guana Bay (St. M. 17), a large and shallow lake is present, bordered at its southern end by coral limestone. The muddy flats are inhabited by numerous fiddler crabs. The outlet of the lagoon is shut off by a sand bar, which appears to be so low, however, that surplus water during heavy rains may run off easily. Along the shore *Batis* and *Conocarpus* form a dense vegetation.

Finally, in the eastern part of Old Building Pond (St. M. 29c and 29d) no algae have been found.

CHAPTER V

SUMMARY AND DISCUSSION

A. SUMMARY

Along the coast of St. Martin, St. Eustatius and Saba the rocks above sea-level often show a number of differently coloured zones. This is clearly visible when the coast over a larger distance is formed by one type of rock, as for instance on Saba. In many places a light-coloured belt is seen above the zone with algal growth; still higher the rocks are dark-coloured (Fig. 11, 17 and 22). The uppermost part of the light-coloured zone indicates rather exactly the height of the waves at low tide (splash zone). The dark-coloured zone is wet from finely dispersed water (spray zone). Sometimes differently coloured zones are formed (Fig. 13) which may, however, be correlated with the above mentioned ones on account of their organisms.

Below the algal zone on open coasts hardly any other zones can be discerned, since the various communities of plants and animals form here a mosaic-like pattern.

On rocky bottoms in the sublittoral region next to the lithophilous algae many corals are present. When the rocks are covered with a thin layer of sand, a rich fauna of Gorgonids may be present; when this layer becomes thicker, the horny corals are replaced by seagrasses and psammophilous algae.

From our data the following facts about the rocky coasts, from the supralittoral margin to the uppermost part of the sublittoral region, are obvious.

A zonation for molluscs is distinctly visible. *Tectarius tuberculatus* and *Littorina ziczac* are found in the spray zone, which is usually recognizable by a dark colour, caused by microscopic bluegreen algae. The rocks in this zone as a rule are almost dry. Chitons are usually present at a somewhat lower level, in places which are constantly washed by the waves; generally they indicate the upper limit of the waves at low tide (e.g. Figs. 17 and 21). They are mostly found in a lightcoloured zone.

In places where the rocks are not exposed to the sun, Chitons may occur at much higher levels than is normally the case, when, however, the rocks are scoured by sand they are scarce or absent. *Nerita tessellata* is common among the Chitons, but may also be found at a lower level among the highest algae.

Under overhanging rocks, and also in other places without direct sunlight a characteristic algal vegetation is found, mainly consisting of *Bostrychia*-species and *Polysiphonia howei;* the Bostrychias at a somewhat higher level than *Polysiphonia*. Protected by overhanging rocks and reached by the waves at high tide *Bostrychia tenella* may form a continuous zone at about 10—20 cm above the water level, with *Polysiphonia howei* immediately below (Great Bay; Fig. 17). In some cases, however, the Bostrychias may be absent (Little Bay near Kay Bay Hill, and near Fort Amsterdam). Other species with a comparable life form may also be found at a high level in shady places, for instance *Lophosiphonia cristata* (Little Bay) or *Loposiphonia* spec. of which very large patches were found on St. Eustatius. Next to these species hardly any algal growth is observed in the littoral region of St. Martin, St. Eustatius and Saba: the high air temperatures and the strong sunshine cause a desiccation of all organisms in the tidal zone. Moreover, because of the oscillations of the mean sea-level during the year, the mortality of the highest algal growth may be considerable.

In the zone, which is also washed by the waves at low tide, a moss-like algal vegetation may be found, with many intermingled species, such as: *Polysiphonia ferulacea, Lophosiphonia cristata, Jania adhaerens, Cladophoropsis membranacea, Cladophora fuliginosa, Centroceras clavulatum, Valonia ocellata, Dilophus alternans, D. guineensis, Hypnea musciformis, H. spinella, Spyridia aculeata, Laurencia intricata, L. papillosa, L. microcladia, L. gemmifera, Caulerpa racemosa, Dictyosphaeria cavernosa, Dictyota* spp., *Padina sanctae-crucis, P. gymnospora, P. vickersiae, Grateloupia cuneifolia, G. filicina, Pocockiella variegata, Ulva fasciata, Gelidiella acerosa, Amphiroa fragilissima, Stichothamnion antillarum, Pterocladia pinnata, Sargassum* spp., and *Wrangelia argus.*

Combination of species may vary from place to place. On flat rocks, for instance beachrock plates, such species as *Polysiphonia ferulacea, Lophosiphonia cristata, Jania adhaerens, Centroceras clavulatum* may be predominant. On protruding rocks, as for instance coral limestone, *Laurencia papillosa* and also *Padina sanctae-crucis* may be abundantly present. This moss-like algal growth may be found in many places along the coast of the islands, particularly in places with strong wave-action.

When the boulders rest in a sandy bottom and much sand is carried by the waves, in several places on the coast of St. Martin a vegetation has developed, consisting of *Chondria tenuissima* and *Digenea simplex*, together with a number of species partly mentioned already when describing the moss-like algal vegetation, for instance *Padina sanctae-crucis, P. vickersiae, Bryothamnion triquetrum, Laurencia microcladia, L. papillosa, L. poitei, Polysiphonia ferulacea*, and *Sargassum* spp.

A clear distinction cannot be made between these two algal vegetations which grow about the same level, as the amount of sand carried by the waves varies from place to place.

A number of algal species is very sensitive to the scouring sand: *Pocockiella variegata* and crustforming Lithothamnia are for instance absent in these places.

A number of the algae present under these circumstances has adapted itself to this special habitat, for instance in developing long trichoblasts, which provide extra protection to the growing tips.

In a few places a distinct zonation of the brown algae is observed, e.g. in Little Bay off Kay Bay Hill. Here zones are present of respectively *Turbinaria turbinata* (in the deeper water), *Sargassum* spec. and *Padina sanctae-crucis* (in the highest places). These species are found on horizontal or gently sloping places. Other kinds of algae, e.g. *Laurencia papillosa* and *L. microcladia* are mostly found on vertical rocks, together with *Pocockiella* and the crustlike Lithothamnia, which form together a mosaic, nearly completely covering the rock. Apparently the slope of the rock surface determines the establishment of these species. Our observations confirm that *Turbinaria* and *Sargassum* prefer exposed places, and *Padina* less exposed habitats.

In many places crustlike Lithothamnia are numerous. On open coasts they may cover almost the entire rock surface below water level. Under overhanging rocks they may be found as high as the Chitons. In exposed places they form rock-covering mosaics together with *Pocockiella;* in protected places Lithothamnia are predominant; both species are absent in the proximity of corals.

A zone of barnacles was observed in a few places. In most cases they are found on the flat upper side of rocks, together with Chitons and the highest algae.

Echinometra lucunter is abundant in the zone of the highest algae, but may also occur at a lower level, to a depth of about 50 cm. In deeper water other species of sea urchins are found, e.g. *Diadema antillarum*, generally in large numbers, both on sandy and on rocky bottoms, and *Lytechinus variegatus*, mostly solitary in seagrass vegetations.

The richest algal vegetation is found on moderately exposed coasts. When the swell is heavy, it tends to impoverish the flora, and the plants become short and compactly built, but even then they may become detached. In quiet water which is insufficiently renewed, the vegetation may be poorly developed too (cf. Guana Bay versus Baie de la Grande Case).

In spite of the fact that the significance of the subsoil for marine organisms is fundamentally different from that for land-plant life, the type of rock may still be of influence on the algal vegetation: the possibilities for attachment may be different; niches or small rock pools may offer special habitat conditions.

On smooth doleritic rocks in Great Bay near Fort Amsterdam, the algal vegetation for nine-tenths consists of *Chaetomorpha media*. Normally this species is firmly attached with a branched system of rhizoids, which is much less the case on smooth doleritic boulders where other algae are almost absent. The rocks of the Point Blanche formation also offer special conditions for attachment (Little Bay off Kay Bay Hill).

When niches are formed, as for instance in coral limestone or the limestone cliffs of the Low Lands formation, favourable conditions are offered for sciophilous algae, such as *Bostrychia* and *Polysiphonia howei.*

In tropical countries the difference in water absorption between porous course-crystalline rocks and fine-crystalline rocks is less reflected in the level in which the algal zones of the littoral region are found than in temperate regions (DEN HARTOG, 1959, p. 29). This is due to the high air temperatures, and fluctuations in mean sea-level.

In the sublittoral region, different communities of plants and animals have developed in a certain relation to the type of bottom. Where the rocky coast forms a horizontal platform at a depth of a few meters, this may harbour several corals and Gorgonids, and also a number of litho-philous algae such as *Amphiroa hancockii, Galaxaura* spp., and *Halimeda discoidea.* In some localities corals, such as *Millepora alcicornis, Eusmilia fastigiata, Porites porites, Acropora cervicornis, A. palmata, Agaricia agari-cites* and *Montastrea* are found up to rather high levels.

Although corals are of general occurrence on St. Martin, St. Eustatius and Saba, the beaches of coral debris, well-known on Bonaire and Curaçao, are not present. This possibly may be explained by the fact that the sea around St. Martin is rather shallow, and the bottom over large areas consists of sand.

When the seabottom consists of sand, the littoral region and the upper-most part of the sublittoral region may be devoid of algae. At a greater depth where the sandy bottom is more stable, vegetations of *Thalassia testudinum* and *Syringodium filiforme* and also of algae may develop. Especially *Thalassia* is important for accumulating sand: the *Thalassia* fields gradually rise and finally may protrude 30—50 cm or more above the surrounding seabottom (cf. Simson Bay and Mouth Piece Bay).

In seagrass vegetations many species of algae are found; particularly re-presentatives of the order Siphonales may be fairly abundant, also in bare sandy bottoms which are somewhat settled. A seagrass vegetation with algae is not only found on open sandy coasts but also in the lagoons, especially at their shallow entrances, where the water is more quiet than in open bays. The presence of numerous spherical-branched Lithothamnia among the *Thalassia* vegetation, at a depht of about 40 cm, near the entrance of Oyster Pond, is remarkable.

The algal vegetations on the roots of *Rhizophora* in Oyster Pond and in the entrance of Fish Pond are exactly as described in the literature. In the littoral region *Bostrychia binderi* and *Polysiphonia howei* are common; in the uppermost part of the sublittoral region many other species are present, especially *Acanthophora spicifera,* together with *Caloglossa leprieurii, Murrayella periclados* and *Ulva lactuca.*

In the rear part of the lagoons, and along the banks of lagoons completely shut off from the sea, the bottom is muddy and the salt content shows remarkable fluctuations. Only a few euryhalinic algal species are found in these shallow pools; especially *Batophora oerstedi* and *Enteromorpha* spp., in most cases attached to coral fragments, pieces of wood or shells.

B. DISCUSSION

From the data for St. Martin, St. Eustatius and Saba it follows that, generally speaking, our observations confirm the published information on the Caribbean area. This particularly applies to the steep cliffs of coral limestone; nearly all the literature bears upon this kind of rock. The same general features, however, may be observed on parts of the coast consisting of other rock formations.

The s u p r a l i t t o r a l   m a r g i n is clearly indicated by the snails *Littorina ziczac, Tectarius tuberculatus* and several species of *Nerita.*

This zone has been called "Littorinoid zone" (Voss & Voss, 1955; RODRIGUEZ, 1959) or "Littorine community" (NEWELL et al., 1959), and "spray zone" (the zone "from the M.H.W.S. up"; RODRIGUEZ, 1959).

The STEPHENSONs and NEWELL et al. divide the supralittoral margin into three zones, from above downward indicated as the white, grey and black zone. These zones show a distinct vertical distribution of the molluscs, especially of *Nerita.*

This subdivision with regard to differences in colour of the rocks cannot be applied to St. Martin, St. Eustatius and Saba; on the other hand differences in the vertical distribution of the *Nerita*-species were observed too, most conspicuously on gently dipping rocks.

The   e u l i t t o r a l   r e g i o n is recognizable by the presence of Chitons, barnacles and several species of gastropods.

Locally in this zone, in more or less shady places, a few algae have been found, viz. *Bostrychia binderi, B. tenella, Polysiphonia howei,* and in some places *Lophosiphonia cristata.* These algae, and also the Chitons and barnacles, grow high in the eulittoral region. In the lower part crustlike Lithothamnia may cover large areas of the rock, especially in places not directly exposed to sunlight. The number of barnacles in the eulittoral region may vary considerably, but is generally small.

This zone is called "midlittoral zone" by the STEPHENSONs, and "Balanoid zone" by Voss & Voss. RODRIGUEZ uses the same term but also speaks of "splash zone", the zone "from M.H.W.S. down". T. A. & A. STEPHENSON (1950), and NEWELL et al. (1959), distinguish an upper yellow and a lower yellow zone, respectively characterized by *Chthamalus stellatus, Bostrychia* and *Polysiphonia, Spiroglyphus irregularis* and *Valonia*

76

ocellata. LEWIS (1960) even discerns three zones, the black, green, and pink zone. The black zone is characterized by *Bostrychia tenella* and *Polysiphonia howei,* often forming dense vegetations; in the green zone especially several Chitons and *Thais* species are observed. The pink zone owes its name to pink-coloured Lithothamnia, which cover the rock to about high water mark. In fact the division of LEWIS comes close to that of the STEPHENSONS. Therefore at least two zones may be distinguished in the eulittoral of the Caribbean.

The richest development of algae is found in the s u b l i t t o r a l r e g i o n. Mostly in its uppermost part a remarkable algal vegetation of short, closely branched, moss-like plants has developed. Most species of this vegetation may also be observed in deeper, more quiet water, where they are much better developed. A number of species is found exclusively in exposed places. Very abundant at the same level is the urchin *Echinometra lucunter,* which has its lower limit at a depth of 45—50 cm. When a lot of sand is carried by the waves, species such as *Chondria tenuissima* and *Digenea simplex* are abundant. Much coral, especially *Millepora alcicornis,* may be found as high as the uppermost part of the sublittoral region.

T. A. & A. STEPHENSON (1950) are of the opinion that this uppermost part of the sublittoral region should be regarded as a separate zone, "infralittoral fringe" (for the Florida Keys they speak of a "lower platform"). VOSS & VOSS (1955) and also RODRIGUEZ (1959) call this zone "*Echinometra*-zone". NEWELL et al. (1959) however, stress the occurrence of other organisms and they distinguish a "*Millepora* community" in the uppermost part of the sublittoral region. LEWIS (1960) uses the term "surfzone".

Although various authors emphasize different aspects, their observations show a certain similarity. Locally, however, there may be variations; NEWELL et al. (p. 210) point to a number of similarities and differences between their "*Millepora* community" and the "lower platform" described by the STEPHENSONS; they are of the opinion that the points of difference at least are of equal importance as the points of resemblance.

RODRIGUEZ describes for Margarita a number of algal vegetations in relation to the degree of exposure and the absence or presence of tidal currents. These differences, connected with local factors, will be discussed later.

From the foregoing it is evident that in the uppermost part of the sublittoral region a number of remarkable combinations of plants and animals may occur, such as *Laurencia papillosa* and *Valonia ocellata,* both numerous, and *Cladophoropsis membranacea, Centroceras clavulatum, Ceramium byssoideum, Herposiphonia* ssp., *Jania capillaca, Polysiphonia ferulacea* and *Spyridia filamentosa,* all together forming a thick and interwoven carpet. *Echinometra lucunter* is also characteristic.

The lower limit of many species, which are abundant in the upper part of the sublittoral region, may be variable. Many of them also grow in deeper and quiet water where they are often much better developed. A number of them, however, has a distinct lower limit: *Turbinaria* occupies the splash-zone (only once a small plant was observed at a depth of 50 cm); *Echinometra* occurs to a depth of 50 cm, which is definitely lower than the lower limit of *Turbinaria; Millepora alcicornis* again has a still lower limit.

DEN HARTOG (1959) considers the distinction of a separate sublittoral margin superfluous, since it refers only to the uppermost part of the sublittoral region. However, at this level a definite and separate community may be found, differing from the communities at a somewhat lower level, which are constantly submerged. Outside the Caribbean area, this was observed in several other parts of the world. For the Gold Coast LAWSON (1957) reports *Sargassum vulgare* and *Dictyopteris delicatula* in a definite and narrow zone. Both species also occur in the Caribbean, but are not limited to such a narrow zone. A separate sublittoral margin is also described by GUILER (1953, 1955) for Tasmania and by WOMERSLEY & EDMONDS (1958) for the southern coast of Australia.

In the continuously submerged part of the sublittoral region, depending on the nature of the bottom, a mosaic of communities has developed. A summary of these communities is given above (Chapter V A, p. 75).

Not all communities described by NEWELL et al. (1959) for rocky bottoms and sandy or muddy bottoms could be distinguished on St. Martin, St. Eustatius and Saba, our observations being too scarce. The main features, however, could be confirmed. In this connection attention may be drawn to the detailed description of algal habitats by TAYLOR (1960), which presents an accurate picture of the algal vegetation in the western, tropical part of the Atlantic Ocean.

When comparing our observations with published data on the Caribbean, it is obvious that the observations may be strongly influenced by all kinds of local factors.

VOSS & VOSS (1955) describe a "Porites-coralline zone" in water of 30—60 cm depth along the eastern shore of Soldier Key. It comprises a rather dense growth of unattached forms of coralline algae such as *Jania*, *Amphiroa* and *Goniolithon*, intermixed with low and scattered *Porites furcata*. These low beds are not continuous but may be interrupted by *Thalassia* and *Syringodium*. T. A. & A. STEPHENSON (1950) did mention such patches of coralline algae of the *Goniolithon strictum* type for the Florida Keys, but did not name them. RODRIGUEZ (1959) could not find this zone in Margarita.

On St. Martin the same community was found in Oyster Pond, in water of about 40 cm deep. Comparable observations were made on Bonaire (Lac; ZANEVELD, 1958) and on Curaçao (Awa di Oostpunt). This combination of species is found always under the same circumstances: shallow water with *Thalassia* and little wave action. On Soldier Key these requirements are found over large areas, in other places only locally.

For Margarita island RODRIGUEZ (1959) describes a number of algal communities, the composition of which is closely related to the degree of exposure. He states, however, that other authors did not report the replacement of *Ulva fasciata* by *Grateloupia cuneifolia* in places with strong tidal currents. Therefore it seems that his communities are only of local significance.

When localities are situated close together, differences may still be observed between the algal vegetations, which cannot always be explained; species may be nearly absent when the habitat might be regarded as suitable. FELDMANN & LAMI (1936) did not succeed in collecting the red alga *Catenelia opuntia* on Guadeloupe, in spite of intensive searching for it (though WAGENAAR HUMMELINCK found the same species in 1964 on the mangroves of the Rivière Salée). Apparently this species is not common there, as is the case on St. Martin.

Several authors went into a discussion as to how the zones and combinations of organisms should be interpreted (a.o. BRAUN-BLANQUET, 1964; FELDMANN, 1951; DEN HARTOG, 1959; NEWELL et al., 1959; RODRIGUEZ, 1959; SKOTTSBERG, 1941 and WOMERSLEY & EDMONDS, 1958). Several attempts were also made to arrive at a clear terminology.

Marine and terrestrial communities show fundamental differences. In the sea, the substrate in the first place is of importance for attachment and should not be seen as a habitat in a physiological sense.

Several authors call attention to the fact that in the littoral and sublittoral region all organisms grow together, i.e. plants and animals form one integrated unity (SKOTTSBERG, 1941; T. A. & A. STEPHENSON, 1949; RODRIGUEZ, 1959; BRAUN-BLANQUET, 1964). In this connection BRAUN-BLANQUET makes the following remarks. Below the water level "leben Pflanze und Tiere in derart enger Verbindung, dass er von vornherein angezeigt erscheint, die alle Lebewesen umfassende "Biozönose" in ihrer Gesamtheit als Einheit nach den pflanzensoziologischen Methoden der Zürich-Montpellier-Schule aufzunehmen. Die Mittberücksichtigung der Tiere verlangt allerdings eine Anpassung des Aufnahmeverfahrens".

However, little has been investigated along these lines. Only by teamwork can a better insight be obtained into the laws governing the distribution of animals and plants. TAYLOR (1960) in his paragraph on "Algal habitats" points to the often superficial and faulty descriptions in many publications, which must be ascribed to the fact that it is hardly possible for one man to survey everything at once.

The distinction of communities of algae is far from easy. Sometimes, a few species may form a community, characterized by one or a few dominant species. Often, however, it appears that, in spite of the fact that circumstances seem to be equal, large variations occur in algal vegetations. Even 20 or 30 species may grow intermingled in various proportions. The limitation of communities then appears very difficult or even impossible, particularly in the sublittoral region (SKOTTSBERG, 1941).

In the sublittoral region investigations were set up by MOLINIER (1960) according to the method of BRAUN-BLANQUET; he worked in the western part of the Mediterranean, and gave a description of the sublittoral communities of Cap Corse (Corsica). For the algal communities of the coastal line ("Uferhaftergesellschaften") matters are easier and more literature is at our disposal. BRAUN-BLANQUET (1964) mentions the work of KORNÁS et al. (1959/60) for the Danziger Busen, and DEN HARTOG (1959) for the epilithic algal communities along the coast of the Netherlands.

In the publications mentioned above several formations, associations and sociations are described. This type of research, however, has been too incidental to allow generalization, but it offers perspectives for the future.

For the epilithical algal communities DEN HARTOG (1959) distinguishes several sociations and associations, which are united into a number of formations on the basis of physiognomy and stratification. He follows WESTHOFF (1951) in definiting sociation and association.

S o c i a t i o n : community containing at least one constant species and consisting of one or more layers with one dominant in each layer.

A s s o c i a t i o n : community with a more or less constant floristic composition and with a specific assemblage (i.e. the characteristic species joined with the constant companion species of the community).

In practice this means for DEN HARTOG that the term sociation is applied to vegetation-units not sufficiently characterized by one or more dominant species and also without characteristic species, but nevertheless taking a well-defined place in the zonation. The only essential difference between sociation and association is the absence of characteristic species in the sociations and their presence in the associations.

DEN HARTOG (1959) unites the different communities in entities of a higher rank. In vegetational science, communities are usually united to alliances, but since little is known on this point for algal communities preference is given to a classification in formations, based on physiognomy and stratification, partly also on their place in succession and zonation. For an algal formation DEN HARTOG (1959) gives the following definition (a modification of a description given by KJELLMANN in 1878): "An algal formation is a part of the algal vegetation whose aspect is determined by the dominance of a definite life form or by the joint dominance of some such life forms" (a summary of the ideas of DEN HARTOG is found on p. 99—100 of his publication).

NEWELL et al. (1959) also go into the purport they give to the terms used in their paper on the Great Bahama Bank dealing with the relation between the nature of the bottom and the communities of organisms. With the term community these authors try to include a wide entity; the term is used to indicate any defined group of organisms living together. They distinguish three types of communities: "habitat-community", "organism-community" and "biocoenosis". "The habitat community is a natural association of organisms set apart according to certain defined features of the environment", for example, "in the Bahamas we can recognize as the rock-bottom community all those benthic organisms that inhabit rock bottom". "The rock-bottom community can, in turn, be subdivided into the coral-reef community, the intertidal rocky-shore community, the community of infratidal rocky prominences and the rock pavement community".

According to NEWELL et al. "organism communities" may be distinguished on the basis of the common occurrence of the organisms, without paying attention to the habitat. "In the Bahamas appropriately defined habitat communities show close correspondence with organism communities. Thus our experience supports the general conclusion of nearly all students of marine bottom communities that the character and distribution of the communities are closely correlated with the substrate character". "This correlation is, however, never exact, and in some cases there is little or no evident correlation".

"Both habitat and organism communities essentially are empirical statistical concepts in which the community is viewed as the simple sum of all the organisms living in a particular place without regard to any of the complex interspecies relationships that may exist."

"In a third, or dynamic, concept of ecological communities, emphasis is given to the mutual dependence of animals and plants which not merely exist in geographic proximity but are bound together by their ecological functions. For this sort of community, the term "biocoenosis" is properly employed. This dynamic approach is not suited to regional studies, such as this investigation of the Great Bahama Bank and it rarely is applicable to palaeontologic work".

The term "zone" is frequently used in all marine-biological publications. Usually a broad meaning is given to this term indicating a belt or strip along the coast where definite organisms may be present. Also differently coloured bands along the coast may be referred to as zones (for instance by T. A. & A. STEPHENSON, 1950, and by LEWIS, 1960). The STEPHENSONS (1949) use the word zone also to indicate the different coastal belts. DEN HARTOG (1959), however, objected to this meaning of the term and prefers the word "region" in that case.

In our review on zonation in the Caribbean we have tried to apply the same terminology as DEN HARTOG (1959). Our use of littoral region, sublittoral region, etc. is different from that used by the STEPHENSONS (1949, 1950). The word margin is used instead of the term fringe proposed by them (see DEN HARTOG 1959, p. 82).

As the main purpose of our investigations was an inventarisation of the marine algae found on the Netherlands Antilles, our own observations and notes in many cases appear to be too incomplete to present clearcut descriptions of communities of plants (and animals).

From the descriptions given of the algal vegetations in different sampling places, nevertheless a number of combinations of organisms may be compiled. Several of these combinations have been described, e.g. the "*Bostrychietum*" on the roots of mangroves and the growths of *Sargassum* and *Turbinaria* in exposed places. With present knowledge is seems most convenient to indicate these combinations of organisms with more or less neutral terms, as was done by other investigators of the marine flora and fauna in the Caribbean area.

When in some cases we speak of communities, this should be interpreted as organism communities, in some cases perhaps as habitat communities in the sense of Newell et al. (1959). Good examples are the vegetations of *Chondria littoralis* and *Digenea simplex* on rocks constantly washed by waves loaded with sand, of *Chaetomorpha media* near the low water mark, of several representatives of the order Siphonales in seagrass vegetations, the vegetations formed nearly exclusively of *Acanthophora spicifera* on the roots of mangroves in the sublittoral region, and the growths of *Rhipocephalus phoenix* in the gullies behind the beachrock formations a.o.

For the rest is does not help us to provide all combinations of organisms with the ending -etum. Also T. A. & A. Stephenson (1949, p. 302) point to this fact and give the following quotation from Singer: "From the beginning, however, it (ecology) has been cursed, more than most sciences, by a horde of technical terms, equally hideous, unnecessary and obfuscating".

For the sublittoral region it is better not to speak of zones as is done for instance by Voss & Voss (1955) and Rodriguez (1959). A neutral and probably better general term could be the word "beds". In the sublittoral region usually a number of clearly recognizable combinations of organisms may be found, which together form a mosaic closely related to the type of bottom. This has been pointed out several times, both in discussing the literature and our own observations. Distinctly recognizable are a formation of seagrasses and green algae of the order Siphonales, and a mangrove formation, on sandy bottoms. On rocky subsoil a formation of corals and horny corals is found and also extensive vegetations of lithophilous algae. Sedentary animals thus may strongly influence the aspect. If communities of plants and animals are to be distinguished within those formations, much attention has to be paid to animal organisms and to the composition of the bottom.

CHAPTER VI

GEOGRAPHICAL REMARKS

The first survey of the algae of the Caribbean area is published by MURRAY (1889). During the preceding years several important papers had been issued, e.g. for Cuba (MONTAGNE, 1842), Mexico (J. G. AGARDH, 1847), Florida (W. H. HARVEY, 1852/58) and Guadeloupe (MAZÉ & SCHRAMM, 1870/77), and since MURRAY uses the work of MAZÉ & SCHRAMM as an important source of information, he portrays the West Indian flora with more endemic species than seems justified at present. COLLINS (1901) in his paper on the marine algae of Jamaica also pays some attention to plant-geographical problems.

Several papers have appeared since on the marine algae of various parts of the Caribbean, e.g. by BØRGESEN (1913—20) for the Virgin Islands, by HOWE (1920) for the Bahamas, and by COLLINS & HERVEY (1917) for Bermuda. At the end of his work on the Virgin Islands BØRGESEN (1920) compares the West Indian algal flora with that of the other side of the Atlantic Ocean and the Mediterranean Sea, and also with that of the Indo-Pacific Ocean.

From the table in which he summarises his results (Table 6) it is clear, probably unexpectedly, that there is a remarkable resemblance between the algal floras mentioned. Especially for the green algae there is a great number of species in common.

Table 6

THE ALGAL FLORA OF THE VIRGIN ISLANDS
compared with that of other parts of the world (BØRGESEN, 1920)

	Total number of species found	Species found in the West Indies and surrounding seas only	Species hitherto only found at the islands in question	West Indian species also found in the Mediterranean Sea and adjacent parts of the Atlantic	West Indian species also found in the Indo-Pacific Ocean
Chlorophyceae	90	33	11	35	46
Phaeophyceae	45	20	5	14	18
Rhodophyceae	192	108	46	63	47
	327	161	62	112	111

MURRAY (1889) reports this fact already. He thinks of a transport via the Cape of Good Hope. BØRGESEN (p. 495), however, assumes a connection between the Pacific and the Atlantic in Tertiary time, and he summarizes his conclusions as follows: "The algal flora of the West Indian Islands in question (Virgin Islands) shows a strikingly great resemblance to that of the Indo-Pacific Ocean. This applies especially to certain, undoubtedly very old, groups of Chlorophyceae. The Rhodophyceae, on the other hand, show less resemblance to those from the Indo-Pacific Ocean, being more closely related to the algal flora occurring in the Mediterranean — Atlantic territory" — "The great similarity between those two algal floras: the West Indian and the Indo-Pacific, which in our days are so distinctly separated, had its natural explanation in a prehistoric old connection between the two oceans."

TAYLOR, who has been actively studying the West Indian algal flora from 1926 onward, expresses his ideas on the distribution of species in his papers of 1950 and 1955. In his Flora of 1960 only his earlier conclusions are repeated. There are some 790 well-defined species known from the area, exclusive of Myxophyceae, diatoms, flagellates, and the like. Of these 790 species only 317 are known from at least 5 major islands or countries, and these species were checked by TAYLOR according to their geographical distribution (Table 7).

Table 7

GEOGRAPHICAL DISTRIBUTION OF CARIBBEAN ALGAE (TAYLOR, 1955)

Number of species	Chloroph. 92	Phaeoph. 53	Rhodoph. 172	All groups 317
Distribution:				
Caribbean, strictly	29.3%	18.8%	30.8%	28.4%
Caribbean and northern	12.0%	5.7%	12.2%	11.0%
Caribbean and southern	34.8%	47.2%	27.9%	33.1%
Caribbean and widespread	23.9%	28.3%	29.1%	27.5%

From this table it is clear that the Caribbean is situated far north of the equator; owing to the Gulfstream its algal flora extends to Florida (30° N.L.) and Bermuda (32° N.L.). Towards the south the Caribbean species occur as far as the southern border of Brazil (Brazil current), near Uruguay (35° S.L.) many of them are replaced by those of a temperate element.

The Caribbean algal flora deserves its name only because the Caribbean

Sea is the area of its greatest diversity in forms. The flora extends only slightly northwards (1000 km), but goes much further southwards (3000 km). Consequently, it may be preferable to speak of a West Atlantic tropical flora. Comparison of this flora with areas far afield is rather doubtful in the present state of knowledge.

TAYLOR (1955) concludes „that the rich Caribbean flora has a high proportion of pantropical and subtropical algae, some relation to the eastern Atlantic, less than earlier has been suggested to the floras of the Indian and Pacific Ocean, and a marked individuality of its own."

In 1961 TAYLOR also deals with their vertical distribution. His conclusions are in the first place ecological and of little importance to plant geography. In deeper water especially species are met with which are shade-tolerant and adapted to anchorage in soft sandy or muddy bottoms, and to relatively low temperatures. Otherwise these species are not all obligatory sciophytes: many *Caulerpa* species are found below 100 m of depth.

From our own observations no far-reaching conclusions in the field of plant geography may be expected. The investigation of many localities supplements our knowledge of the distribution of Caribbean species, but only one single new species has been found in hundreds of samples: *Stichothamnion antillarum* (VROMAN, 1967).

Most species have an even distribution between Cape Canaveral and Rio de Janeiro, but at the same time the composition of the communities of plants and animals also may be mutually compared which means a considerable simplification when comparing the data from different parts of the Caribbean.

To gain a better knowledge of the West Indian algal flora — besides our own material — a large number of the samples collected by dr. P. WAGENAAR HUMMELINCK was also studied. The results are summarized in Table 8. A survey of these data for each collecting place is given separately in Chapter VII.

The material has been deposited in the Herbarium of the Vrije Universiteit (Free University) in Amsterdam, with the exception of some fragmentary material of well-known and common species. Duplicates of a great number of species were sent to the
Herbarium P. et H. Huvé, Marseille, France;
Herbarium Smithsonian Institution, U.S.N.M., Washington, D.C., U.S.A.;
Herbarium H. J. Humm, Duke University, Durham, N.C., U.S.A.;
Herbarium W. R. Taylor, University of Michigan, Ann Arbor, U.S.A.;
Rijksherbarium, Leiden, Netherlands;
Farlow Herbarium, Harvard University, Cambridge, Mass., U.S.A.;
Caraïbisch Marien-Biologisch Instituut, Curaçao, N.A.;
Herbarium M. Díaz-Piferrer, University of Puerto Rico, Mayagüez, P.R.
Herbarium M. S. Doty, Honolulu, Hawaii.

Table 8

mentioned in 'List of localities and species' (Chapter VII)

Species	Anguilla	Barbuda	Islote Aves	Saba	St. Barts	St. Croix	St. Eustatius	St. John	St. Kitts	St. Martin
CHLOROPHYTA										
Acetabularia calyculus Quoi & Gaimard	—	—	—	—	—	—	—	—	—	×
Acetabularia crenulata Lamouroux	—	×	—	—	—	—	—	—	—	×
Acetabularia polyphysoides Crouan	—	—	—	—	—	—	—	—	—	×
Anadyomene stellata (Wulfen) C. Agardh	—	×	—	—	—	—	—	—	—	×
Avrainvillea longicaulis (Kützing) Murray & Boodle	—	—	—	—	—	—	—	—	—	×
Avrainvillea rawsoni (Dickie) Howe	—	×	—	—	—	—	—	—	—	×
Batophora oerstedi J. Agardh	—	×	—	—	—	×	—	—	—	×
Blastophysa rhizopus Reinke	—	—	—	—	—	—	—	—	—	×
Boodlea composita (Harvey & Hooker fil.) Brand	—	—	—	—	—	—	—	—	—	×
Boodleopsis pusilla (Collins) Taylor, Joly & Bernatowicz	—	×	—	—	—	—	—	—	—	?
Bryopsis pennata Lamouroux	—	—	×	—	×	—	—	—	—	×
Bryopsis plumosa (Hudson) C. Agardh	—	—	—	—	—	—	—	—	×	—
Bryopsis spec.	—	—	—	—	—	—	—	—	—	×
Caulerpa ambigua Okamura	—	—	—	—	×	—	×	—	×	×
Caulerpa cupressoides (West) C. Agardh	—	—	—	—	—	—	—	—	—	×
Caulerpa crassifolia (C. Agardh) J. Agardh f. *mexicana* (Sonder) Weber-van Bosse	—	—	—	—	—	—	—	—	—	×
Caulerpa microphysa (Weber-van Bosse) Feldmann	—	×	—	—	—	—	×	—	—	—
Caulerpa prolifera (Forsskål) Lamouroux	—	—	—	—	—	—	—	—	—	×
Caulerpa racemosa (Forsskål) J. Agardh	—	×	—	—	—	—	×	—	—	×
Caulerpa sertularioides (Gmelin) Howe	—	×	—	—	—	—	×	—	×	×
Caulerpa taxifolia (Vahl) C. Agardh	—	—	—	—	—	—	—	—	—	×
Caulerpa verticillata J. Agardh	—	×	—	—	—	—	—	—	—	—
Chaetomorpha aerea (Dillwyn) Kützing	—	—	—	—	—	—	—	—	—	×
Chaetomorpha brachygona Harvey	—	—	—	×	—	—	×	—	—	×
Chaetomorpha clavata (C. Agardh) Kützing	—	—	—	—	—	—	×	—	—	×
Chaetomorpha crassa (C. Agardh) Kützing	—	—	—	—	—	—	—	—	—	×

Species	Anguilla	Barbuda	Islote Aves	Saba	St. Barts	St. Croix	St. Eustatius	St. John	St. Kitts	St. Martin
CHLOROPHYTA (continued)										
Chaetomorpha gracilis Kützing	×	—	—	—	—	—	—	—	—	×
Chaetomorpha linum (O. F. Müller) Kützing	—	—	—	—	—	—	—	—	—	×
Chaetomorpha media (C. Agardh) Kützing	—	—	—	×	—	—	×	—	×	×
Cladophora crispula Vickers	—	—	—	—	—	—	—	—	—	×
Cladophora fascicularis (Mertens) Kützing	—	—	—	—	—	—	—	—	—	×
Cladophora fuliginosa Kützing	—	×	—	—	—	—	—	—	—	×
Cladophora luteola Harvey	—	—	—	—	—	—	—	—	—	×
Cladophora uncinata Børgesen	—	—	—	—	—	—	—	—	—	×
Cladophora spec.	—	—	—	×	—	—	—	—	—	×
Cladophoropsis membranacea (C. Agardh) Børgesen	—	—	—	—	—	×	×	×	×	×
Codium isthmocladum Vickers	—	—	—	—	—	—	—	—	—	×
Codium taylori Silva	—	—	—	—	—	—	—	—	—	×
Codium spec.	—	—	—	—	—	—	—	—	—	×
Dictyosphaeria cavernosa (Forsskål) Børgesen	—	×	—	—	—	—	—	—	—	×
Dictyosphaeria vanbosseae Børgesen	—	—	—	—	—	—	×	—	—	×
Diplochaete solitaria Collins	—	×	—	×	—	—	—	—	×	×
Enteromorpha chaetomorphoides Børgesen	—	—	—	—	—	—	—	—	—	×
Enteromorpha compressa (Linnaeus) Greville	—	—	—	—	—	—	—	—	—	×
Enteromorpha erecta (Lyngbye) J. Agardh	—	—	—	—	—	—	—	—	—	×
Enteromorpha flexuosa (Wulfen) J. Agardh	—	—	—	—	—	—	×	—	—	×
Enteromorpha intestinalis (Linnaeus) Link	—	—	—	—	—	—	—	—	—	×
Enteromorpha lingulata J. Agardh	—	—	—	—	—	—	×	—	—	×
Enteromorpha prolifera (Müller) J. Agardh	—	—	—	—	—	—	—	—	—	×
Ernodesmis verticillata (Kützing) Børgesen	—	—	—	—	—	—	—	—	×	×
Halimeda discoidea Decaisne	—	—	—	—	—	—	—	—	—	×
Halimeda incrassata (Ellis & Solander) Lamouroux	×	—	—	—	—	×	—	—	—	×
Halimeda opuntia (Linnaeus) Lamouroux	×	×	×	—	×	×	—	×	—	×

87

Species	Anguilla	Barbuda	Islote Aves	Saba	St. Barts	St. Croix	St. Eustatius	St. John	St. Kitts	St. Martin
CHLOROPHYTA (continued)										
Halimeda tuna (Ellis & Solander) Lamouroux	—	—	—	—	—	—	×	—	—	—
Microdictyon boergesenii Setchell	—	×	—	—	—	—	—	—	—	—
Neomeris annulata Dickie	—	—	—	—	—	—	×	—	—	×
Neomeris mucosa Howe	—	—	—	—	—	—	—	—	—	×
Penicillus capitatus Lamarck	×	×	—	—	—	×	—	—	—	×
Penicillus dumetosus (Lamouroux) Blainville	×	×	—	—	—	—	—	—	—	×
Penicillus lamourouxii Decaisne	—	—	—	—	—	—	—	—	—	×
Penicillus pyriformis A. & E. S. Gepp	—	—	—	—	—	—	—	—	—	×
Rhipocephalus phoenix (Ellis & Solander) Kützing	—	—	—	—	—	—	—	—	—	×
Rhizoclonium kerneri Stockmayer	—	—	—	—	—	—	—	—	—	×
Siphonocladus rigidus Howe	—	×	—	—	—	—	—	—	—	—
Struvea anastomosans (Harvey) Piccone	—	—	—	—	—	—	×	—	—	×
Udotea conglutinata (Ellis & Solander) Lamouroux	—	—	—	—	—	—	—	—	—	×
Udotea flabellum (Ellis & Solander) Lamouroux	×	—	—	—	—	—	×	—	—	×
Udotea sublittoralis Taylor	—	—	—	—	—	—	—	—	—	×
Ulva fasciata Delile	—	—	—	×	—	—	×	—	—	—
Ulva lactuca Linnaeus	—	—	—	—	—	—	×	—	—	×
Ulvella lens Crouan	—	—	—	—	—	—	—	—	—	×
Valonia aegagropila C. Agardh	—	×	—	—	—	—	—	—	—	×
Valonia macrophysa Kützing	—	—	—	—	—	—	—	—	—	×
Valonia ocellata Howe	—	×	—	?	—	—	×	—	—	×
Valonia utricularis C. Agardh	—	?	—	—	—	—	—	—	—	—
Valonia ventricosa J. Agardh	×	×	—	—	×	—	×	×	×	×
PHAEOPHYTA										
Colpomenia sinuosa (Roth) Derbès & Solier	—	—	—	×	×	—	—	—	—	×
Dictyopteris delicatula Lamouroux	×	×	×	×	×	—	×	—	—	×
Dictyopteris justii Lamouroux	—	×	—	—	—	—	×	—	—	—
Dictyopteris plagiogramma (Montagne) Vickers	—	×	—	—	—	—	—	—	—	×
Dictyota bartayresii Lamouroux	—	—	—	—	—	—	—	—	—	×
Dictyota cervicornis Kützing	—	×	—	×	—	—	×	×	—	×
Dictyota ciliolata Kützing	×	—	—	×	—	—	×	×	×	×

Species	Anguilla	Barbuda	Islote Aves	Saba	St. Barts	St. Croix	St. Eustatius	St. John	St. Kitts	St. Martin
Dictyota dentata Lamouroux	×	×	—	—	—	—	×	—	—	×
Dictyota dichotoma (Hudson) Lamouroux	—	—	—	—	—	—	—	—	—	×
Dictyota divaricata Lamouroux	—	—	—	—	—	—	—	—	×	×
Dictyota indica Sonder in Kützing	—	—	—	—	—	—	—	—	—	×
Dictyota jamaicensis Taylor	—	—	—	×	—	—	×	×	—	×
Dilophus alternans J. Agardh	×	×	—	—	—	—	×	—	—	×
Dilophus guineensis (Kützing) J. Agardh	—	—	—	×	—	—	×	—	×	×
Ectocarpus breviarticulatus J. Agardh	—	—	—	—	—	—	—	—	—	×
Ectocarpus confervoides (Roth) Le Jolis	—	—	—	—	—	—	—	—	—	×
Ectocarpus elachistaeformis Heydrich	—	—	—	—	—	—	—	—	—	×
Giffordia duchassaigniana (Grunow) Taylor	—	—	—	—	—	—	—	—	—	×
Giffordia mitchellae (Harvey) Hamel	—	—	—	×	—	—	×	—	—	×
Giffordia rallsiae (Vickers) Taylor	—	—	—	—	—	—	—	—	—	×
Padina gymnospora (Kützing) Vickers	—	—	—	×	—	—	—	×	—	×
Padina sanctae-crucis Børgesen	×	×	—	—	—	—	—	×	×	×
Padina vickersiae Hoyt	—	—	—	×	×	—	?	—	—	×
Pocockiella variegata (Lamouroux) Papenfuss	—	×	—	×	—	—	×	×	—	×
Sargassum cf. *filipendula* C. Agardh	—	—	—	—	—	—	—	—	—	×
Sargassum cf. *fluitans* Børgesen	—	—	×	—	—	—	—	—	—	—
Sargassum natans (Linnaeus) J. Meyen	—	—	×	—	—	—	×	—	—	×
Sargassum platycarpum Montagne	×	—	—	×	—	—	×	—	—	×
Sargassum polyceratium Montagne	×	×	—	×	—	—	×	×	—	×
Sargassum rigidulum Kützing	—	—	—	—	—	—	—	—	—	×
Sargassum vulgare C. Agardh	—	—	—	?	—	—	?	—	—	×
Sargassum spec.	—	—	—	×	×	—	×	×	—	×
Sphacelaria tribuloides Meneghini	—	—	—	—	—	—	×	—	—	×
Stypopodium zonale (Lamouroux) Papenfuss	—	—	—	—	—	—	×	—	—	×
Turbinaria turbinata (Linnaeus) Kuntze	×	×	×	×	—	—	×	×	—	×
RHODOPHYTA										
Acanthophora spicifera (Vahl) Børgesen	×	—	—	—	—	×	—	—	—	×
Acrochaetium seriatum Børgesen	—	—	—	—	—	—	—	—	—	×
Amphiroa fragilissima (Linnaeus) Lamouroux	×	×	—	—	×	—	×	×	×	×
Amphiroa hancockii Taylor	—	—	—	—	—	—	×	—	—	×

Species	Anguilla	Barbuda	Islote Aves	Saba	St. Barts	St. Croix	St. Eustatius	St. John	St. Kitts	St. Martin
Rhodophyta (continued)										
Amphiroa rigida Lamouroux var. *antillana* Børgesen	—	—	—	—	—	—	—	—	—	×
Amphiroa tribulus (Ellis & Solander) Lamouroux	×	—	—	—	—	—	—	—	—	—
Asparagopsis taxiformis (Delile) Collins & Hervey	—	—	—	—	—	—	—	—	—	×
Asterocytis ramosa (Thwaites) Gobi	—	×	—	—	—	—	—	—	—	×
Bostrychia binderi Harvey	—	—	—	—	—	—	—	—	—	×
Bostrychia moritziana (Sonder) J. Agardh	—	—	—	—	—	—	—	—	—	×
Bryothamnion seaforthii (Turner) Kützing	—	—	—	—	—	—	—	—	—	×
Bryothamnion triquetrum (Gmelin) Howe	×	×	—	—	—	—	×	—	—	×
Callithamnion byssoides Arnott in Hooker	—	—	—	—	—	—	×	—	—	—
Callithamnion cordatum Børgesen	—	—	—	—	—	—	—	—	—	×
Caloglossa leprieurii (Montagne) J. Agardh	—	—	—	—	—	—	—	—	—	×
Centroceras clavulatum (C. Agardh) Montagne	×	×	×	×	—	×	×	×	×	×
Ceramium byssoideum Harvey	—	—	×	×	—	—	×	—	—	×
Ceramium cruciatum Collins & Hervey	—	—	—	—	—	—	—	—	—	×
Ceramium fastigiatum (Roth) Harvey	—	—	—	—	—	—	×	—	—	×
Ceramium floridanum J. Agardh	—	—	—	×	—	—	×	—	—	×
Ceramium nitens (C. Agardh) J. Agardh	—	—	—	—	—	—	—	—	—	×
Champia parvula (C. Agardh) Harvey	—	—	—	×	—	—	×	—	×	×
Chondria atropurpurea Harvey	—	—	—	—	—	×	—	—	—	×
Chondria collinsiana Howe	—	—	—	—	—	—	—	—	—	×
Chondria curvilineata Collins & Hervey	—	×	—	—	—	—	—	—	—	×
Chondria tenuissima (Goodenough & Woodward) C. Agardh	—	—	—	—	—	—	—	—	—	×
Coelothrix irregularis (Harvey) Børgesen	—	—	×	—	—	—	×	×	—	×
Corallina cubensis (Montagne) Kützing	—	×	—	—	—	—	—	—	—	×
Crouania attenuata (Bonnemaison) J. Agardh	—	—	—	×	—	—	—	×	—	×
Crouania pleonospora Taylor	—	—	—	—	—	—	—	—	—	×
Dasya cf.. *corymbifera* J. Agardh	—	—	—	—	—	—	—	—	—	×
Dasya pedicellata (C. Agardh) C. Agardh	—	?	—	—	—	—	—	—	—	×
Dasya rigidula (Kützing) Ardissone	—	×	—	—	—	—	—	—	—	×

Species	Anguilla	Barbuda	Islote Aves	Saba	St. Barts	St. Croix	St. Eustatius	St. John	St. Kitts	St. Martin
Rhodophyta (continued)										
Dasya sertularioides Howe & Taylor	—	—	—	—	—	—	—	—	—	×
Dasya spec.	—	—	—	×	—	—	—	—	—	×
Digenea simplex (Wulfen) J. Agardh	×	×	—	—	—	—	×	—	—	×
Dipterosiphonia dendritica (C. Agardh) Schmitz	—	—	—	—	—	—	×	—	—	×
Enantiocladia duperreyi (C. Agardh) Falkenberg	—	—	—	—	—	—	—	—	—	×
Falkenbergia hillebrandii (Bornet) Falkenberg	—	—	—	—	—	—	—	—	—	×
Galaxaura cylindrica (Ellis & Solander) Lamouroux	—	—	—	—	—	—	×	—	—	—
Galaxaura marginata (Ellis & Solander) Lamouroux	—	—	—	—	—	—	×	—	—	×
Galaxaura oblongata (Ellis & Solander) Lamouroux	—	—	—	×	—	—	×	—	—	—
Galaxaura rugosa (Ellis & Solander) Lamouroux	—	—	—	×	—	—	×	—	—	×
Galaxaura squalida Kjellman	×	—	—	—	—	—	×	—	—	×
Galaxaura subverticillata Kjellman	×	—	—	×	—	—	×	×	—	×
Gelidiella acerosa (Forsskål) Feldmann & Hamel	×	×	—	×	—	—	×	—	—	×
Gelidiopsis intricata (C. Agardh) Vickers	—	—	—	—	—	—	—	—	×	—
Gelidiopsis planicaulis (Taylor) Taylor	—	—	×	—	—	—	—	—	—	—
Gelidium pusillum (Stackhouse) Le Jolis	—	—	—	—	—	—	—	—	×	—
Gracilaria cervicornis (Turner) J. Agardh	—	—	—	—	—	—	—	—	—	×
Gracilaria debilis (Forsskål) Børgesen	×	—	—	—	—	—	—	—	—	×
Gracilaria domingensis Sonder	—	—	—	—	—	—	—	—	—	×
Gracilaria ferox J. Agardh	—	—	—	—	—	—	—	—	—	×
Gracilaria mammillaris (Montagne) Howe	—	—	—	—	×	—	×	—	—	×
Gracilaria verrucosa (Hudson) Papenfuss	—	—	—	—	—	—	—	—	—	×
Gracilaria spec.	—	—	—	—	—	—	×	—	—	×
Grateloupia cuneifolia J. Agardh	—	—	—	×	—	—	×	—	—	×
Grateloupia filicina (Wulfen) C. Agardh	—	—	—	×	—	—	×	—	—	×
Grateloupia spec.	—	—	—	×	—	—	—	—	—	—
Griffithsia tenuis C. Agardh	—	—	—	—	—	—	—	—	—	×
Gymnogongrus tenuis (J. Agardh) J. Agardh	—	—	—	—	—	—	—	—	—	×
Herposiphonia pecten-veneris (Harvey) Falkenberg	—	×	—	—	—	—	—	—	—	×

S p e c i e s	Anguilla	Barbuda	Islote Aves	Saba	St. Barts	St. Croix	St. Eustatius	St. John	St. Kitts	St. Martin
RHODOPHYTA (continued)										
Herposiphonia secunda (C. Agardh) Ambronn	—	—	—	—	—	—	—	—	—	×
Herposiphonia tenella (C. Agardh) Ambronn	—	—	—	—	—	—	×	×	—	×
Heterosiphonia gibbesii (Harvey) Falkenberg	—	—	—	—	—	—	—	—	—	×
Heterosiphonia wurdemanni (Bailey) Falkenberg	—	×	—	—	—	—	—	—	—	×
Hypnea cervicornis J. Agardh	—	×	—	×	—	×	×	—	—	×
Hypnea cornuta (Lamouroux) J. Agardh	—	—	—	—	—	—	—	—	—	×
Hypnea musciformis (Wulfen) Lamouroux	—	×	—	×	—	—	×	—	×	×
Hypnea spinella (C. Agardh) Kützing	—	?	—	×	—	—	—	—	×	×
Hypnea spec.	—	×	—	×	—	—	×	—	—	—
Hypneocolax stellaris Børgesen	—	—	—	×	—	—	—	—	—	×
Jania adhaerens Lamouroux	—	×	—	×	—	—	×	×	—	×
Jania pumila Lamouroux	—	×	—	—	—	—	×	—	—	×
Jania rubens (Linnaeus) Lamouroux	—	—	—	—	—	—	×	—	—	×
Kylinia crassipes (Børgesen) Kylin	—	—	—	×	—	—	—	—	—	×
Laurencia gemmifera Harvey	×	×	—	×	—	—	×	—	—	×
Laurencia intricata Lamouroux	—	—	—	×	—	—	×	—	—	×
Laurencia microcladia Kützing	—	×	—	×	×	—	×	×	—	×
Laurencia obtusa (Hudson) Lamouroux	×	×	×	×	×	—	×	—	—	×
Laurencia papillosa (Forsskål) Greville	×	×	×	×	×	—	×	×	×	×
Laurencia poitei (Lamouroux) Howe	—	—	—	—	—	—	—	—	—	×
Laurencia scoparia J. Agardh	—	—	—	—	—	—	—	—	—	×
Laurencia spec.	—	—	—	×	—	—	×	—	—	×
Liagora ceranoides Lamouroux	—	×	—	—	—	—	—	×	×	×
Liagora decussata Montagne	—	—	—	—	×	—	—	—	—	—
Liagora farinosa Lamouroux	—	—	—	—	—	—	—	×	—	×
Liagora pedicellata Howe	—	—	—	—	—	—	—	—	×	×
Liagora pinnata Harvey	—	—	—	—	—	—	—	×	—	×
Liagora valida Harvey	—	—	—	—	—	—	—	×	—	×
Liagora spec.	—	×	—	—	—	—	—	—	—	×
Lomentaria rawitscheri Joly	—	—	—	×	—	—	—	—	—	—
Lophocladia trichoclados (Mertens) Schmitz	—	×	—	—	—	—	—	—	—	×
Lophosiphonia cristata Falkenberg	—	—	—	×	—	—	×	×	—	×
Lophosiphonia spec.	—	—	—	—	—	—	×	—	—	×
Melobesia farinosa Lamouroux	×	×	—	—	—	×	—	—	×	×

Species	Anguilla	Barbuda	Islote Aves	St. Barts	Saba	St. Croix	St. Eustatius	St. John	St. Kitts	St. Martin
RHODOPHYTA (continued)										
Murrayella periclados (C. Agardh) Schmitz	—	—	—	—	—	—	—	—	—	×
Ochtodes filiformis J. Agardh	—	—	—	—	—	—	×	—	—	×
Peyssonnelia spec.	—	—	—	—	—	—	—	—	—	×
Polysiphonia binneyi Harvey	—	—	—	—	—	—	—	—	—	×
Polysiphonia denudata (Dillwyn) Kützing	—	—	—	—	—	—	—	—	—	×
Polysiphonia ferulacea Suhr	—	×	—	×	×	—	×	×	×	×
Polysiphonia gorgoniae Harvey	—	—	—	—	—	—	—	—	—	×
Polysiphonia cf. *havanensis* Montagne	—	—	—	—	—	—	—	—	—	×
Polysiphonia howei Hollenberg	—	—	—	—	—	—	—	×	×	×
Polysiphonia cf. *subtilissima* Montagne	—	×	—	—	—	—	—	—	—	—
Polysiphonia spec.	—	—	—	—	—	—	×	—	—	×
Pterocladia bartlettii Taylor	—	—	—	—	—	—	×	—	—	—
Pterocladia pinnata (Hudson) Papenfuss	—	—	—	×	—	—	—	—	—	×
Spermothamnion investiens (Crouan) Vickers	—	—	—	—	—	—	×	—	—	×
Spermothamnion spec.	—	—	—	—	—	—	×	—	—	—
Spyridia aculeata (Schimper) Kützing	—	—	—	—	—	—	×	—	—	×
Spyridia filamentosa (Wulfen) Harvey	—	—	—	×	—	×	—	—	—	×
Stichothamnion antillarum Vroman	—	—	—	—	—	—	×	—	—	—
Thuretia borneti Vickers	—	—	—	—	—	—	—	—	—	×
Wrangelia argus Montagne	—	—	—	×	×	—	×	—	×	×
Wurdemannia miniata (Draparnaud) Feldmann & Hamel	—	×	—	—	—	—	×	—	—	—

ANTIGUA

Chlorophyta: *Halimeda opuntia.* — Phaeophyta: *Dictyopteris delicatula, Dictyota cervicornis, Dilophus alternans, Padina vickersiae, Sargassum* spec. — Rhodophyta: *Galaxaura squalida, G. subverticillata, Laurencia papillosa, Melobesia farinosa, Wrangelia argus.*

FOURCHE

Chlorophyta: *Boodlea composita, Chaetomorpha brachygona.* — Phaeophyta: *Dilophus alternans, Padina sanctae-crucis.* — Rhodophyta: *Ceramium byssoideum, Galaxaura rugosa, Jania* spec., *Liagora farinosa.*

NEVIS

Rhodophyta: *Bryothamnion triquetrum, Hypnea musciformis, Laurencia papillosa, Lophocladia trichoclados, Spyridia aculeata.*

ST. THOMAS

Chlorophyta: *Enteromorpha flexuosa, Ulva* spec. - Rhodophyta: *Centroceras clavulatum, Gymnogongrus tenuis.*

The nomenclature in Table 8 is according to TAYLOR (1960), but with a few exceptions. For a number of very polymorphic species BØRGESEN (1913—1920) was followed, e.g. for *Caulerpa* and *Halimeda*. Some species are remarkably constant in form, and therefore the distinction of the species offers no problems; others, however, are very variable. Several authors tend to distinguish many 'small species' in spite of the fact that connecting forms are known. Other investigators prefer to unite all forms to only a few 'large species'. BØRGESEN (1920) thinks that the solution may lie midway. When the material is large enough to show many connecting forms, these should be united to one and the same species. When the material is scarce, one had better be careful and distinguish separate species. Complex species are, e.g. *Caulerpa cupressoides* and *Halimeda incrassata*, which, according to BØRGESEN, comprise several different forms. This in contrary to HOWE (1920) who distinguishes for instance a number of species in the *Halimeda incrassata* complex. BØRGESEN is of opinion "that we get a much better idea of the mutual relationships of the different forms, by connecting those which are obviously related, into larger species, than by dividing them up in a greater number of small ones. This last renders it difficult to form a clear conception of their mutual affinities and equally difficult to compare the geographical conditions of different floras".

The total numbers of marine algae collected on the different islands are represented in Table 9. In this table algae that could not be identified as to the species are not counted if other species belonging to the same genera are listed for the same island; for instance: *Cladophora* spec. is counted for Saba, but not for St. Martin. *Polysiphonia* spec. and *Laurencia* spec. are counted as separate species: of *Polysiphonia* an unnamed species is present (see (BØRGESEN, 1913—1920) and of *Laurencia* a small form occurs which probably may be considered to be an ecological form of *L. papillosa*.

The total number of species (*218*) is considerably lower than that given by BØRGESEN for the Virgin Islands (Table 6: 327). Several reasons may be indicated to explain this difference. Our samples were taken by walking along the coast, or by swimming and diving; they only include specimens from a rather narrow and shallow strip along the shore. BØRGESEN, however, also got samples from a greater depth. He visited the Virgin Islands several times, and was able to pay special attention to difficult groups. Furthermore, much attention was given by him to very small epiphytes, for instance the genus *Acrochaetium*, of which several new species were described.

Our investigation confirms the conclusions by TAYLOR (1955). The Caribbean marine flora is an exceedingly rich one. After elimination of early and ill-described species still 790 well defined species are known. However, there is very little encouragement for the anticipation of discoveries of many new species or rediscoveries of ill-described species

Table 9

TOTAL NUMBERS OF MARINE ALGAE COLLECTED ON THE DIFFERENT ISLANDS
(based on Table 8)

Island	Number of *species*			
	Chloroph.	Phaeoph.	Rhodoph.	Total
Anguilla	7	8	14	29
Antigua	1	5	5	11
Barbuda	20	10	26	56
Fourche	2	2	4	8
Islote Aves	2	4	6	12
Nevis	—	—	5	5
Saba	5	13	31	49
St. Barts	4	4	7	15
St. Croix	5	—	7	12
St. Eustatius	19	17	47	83
St. John	3	8	14	25
St. Kitts	8	4	14	26
St. Martin	64	32	97	193
St. Thomas	1	—	2	3
Total number of *species*	73	34	111	218

(except among the more minute forms). Large parts of the Caribbean area, however, have not been investigated at all or only very fragmentary; many species are known from but one or two reports, although these appear to be reliable. Therefore many new data about the distribution of the species may be brought to light. Our investigation gives a small contribution to that purpose.

CHAPTER VII

LIST OF LOCALITIES AND SPECIES

In the following list of localities, visited by dr. P. WAGENAAR HUMME-
LINCK (station numbers 410—1403) and by the author (numbers 1—35)
the names of the islands are arranged alphabetically.

For the exact situation of the localities see Figures 3 (Saba), 4 (St.
Eustatius) and 10 (St. Martin), and HUMMELINCK's descriptions in *Studies
fauna Curaçao 4*, 1953.

The species are enumerated alphabetically for each algal group (abbreviated to Chl. for Chloropytha, Phae. for Phaeophyta, Rho. for Rhodophyta, and Cya. for Cyanophyta).

Author's names for the species of Chlorophyta, Phaeophyta and Rhodophyta are given in Table 8; those for Cyanophyta may be found in the
publication of KOSTER (1963).

The material of Lithothamnia has not yet been investigated and is not
included in this list.

ANGUILLA

481 FOREST POINT, S. E. coast, 20.VI.1949.

Decaying *Thalassia* with algae, cast on sandy shore.

Phae.: *Padina* spec. — Rho.: *Acanthophora spicifera, Bryothamnion triquetrum, Centroceras clavulatum, Digenea simplex, Laurencia* cf. *obtusa,
Melobesia farinosa.*

1142 SANDY GROUND, N. coast, 19.VI.1949.

Rocky beach with sandy reef; 0—1½ m.

Chl.: *Halimeda incrassata, H. opuntia, Penicillus capitatus, P. dumetosus,
Udotea flabellum, Valonia ventricosa.* — Phae.: *Dictyota ciliolata, D.
dentata, Dictyota* spec.*, Dilophus alternans, Padina sanctae-crucis, Sargassum platycarpum, S. polyceratium, Turbinaria turbinata.* — Rho.: *Amphiroa fragilissima, A. tribulus, Centroceras clavulatum, Digenea simplex,
Galaxaura squalida, G. subverticillata, Gelidiella acerosa, Gracilaria debilis,
Laurencia gemmifera, L.* cf. *obtusa, L. papillosa.* — Cya.: *Dichothrix
fucicola.*

1144 SANDY GROUND, Saltpond, 16.VI.1949.

Ditch between Saltpond and wall; with much *Ruppia* (46 g Cl/l).

Chl.: *Chaetomorpha gracilis.* — Cya.: *Spirulina subsalsa.*

1393 DEEP BAY at Fort Barrington, 17.VII.1955.

Volcanic tuffaceous rock, pebbles, some coarse sand; 0—1 m.

Chl.: *Halimeda opuntia.* — Phae.: *Dictyopteris delicatula, Dictyota cervicornis, Dilophus alternans, Padina vickersiae, Sargassum* spec. — Rho.: *Galaxaura squalida, G. subverticillata, Laurencia papillosa, Melobesia farinosa, Wrangelia argus.*

BARBUDA

1394 MARTELLO TOWER BEACH, S. coast, 8.VII.1955.

Exposed sandy beach with a few boulders; algae collected from pieces of rock; 1—2 m.

Chl.: *Boodleopsis pusilla, Caulerpa microphysa, C. verticillata, Cladophoropsis membranacea, Halimeda opuntia, Microdictyon boergesenii, Penicillus dumetosus.* — Phae.: *Dictyota cervicornis, Dilophus alternans, Padina sanctae-crucis, Sargassum* cf. *polyceratium.* — Rho.: *Amphiroa fragilissima, Bryothamnion triquetrum, Corallina cubensis, Dasya* cf. *pedicellata, D. rigidula, Digenea simplex, Heterosiphonia wurdemanni, Hypnea* spec., *Jania adhaerens, Laurencia obtusa, L. papillosa, Lophocladia trichoclados, Melobesia farinosa, Wurdemannia miniata.*

1395A TWO FEET BAY, E. coast, 10.VII. 1955.

Surfswept limestone cliff; sandy rock pools with few *Thalassia;* 0—½ m.

Chl.: *Cladophora fuliginosa, Cladophoropsis membranacea, Dictyosphaeria cavernosa.* — Phae.: *Dictyopteris delicatula, D. justii, Dictyota dentata, Padina sanctae-crucis, Pocockiella variegata, Turbinaria turbinata.* — Rho.: *Amphiroa fragilissima, Hypnea musciformis, Jania adhaerens, Laurencia microcladia, L. papillosa, Melobesia farinosa, Polysiphonia ferulacea.*

1395B TWO FEET BAY, 10.VII.1955.

Sandy pool; 0—½ m.

Chl.: *Caulerpa sertularioides, Cladophora fuliginosa, Cladophoropsis membranacea, Dictyosphaeria cavernosa, Halimeda opuntia, Penicillus capitatus, Siphonocladus rigidus, Valonia aegagropila, V.* cf. *utricularis, V. ventricosa.* — Phae.: *Dictyopteris delicatula, D. justii.* — Rho.: *Asterocytis ramosa, Digenea simplex, Gelidiella acerosa, Jania adhaerens, Melobesia farinosa.*

1395C TWO FEET BAY, 10.VII.1955.

Submerged shallow ridge; 0—½ m.

Chl.: *Caulerpa racemosa, Cladophora fuliginosa, Cladophoropsis membranacea, Dictyosphaeria cavernosa, Halimeda opuntia, Valonia aegagropila, V. ocellata.* — Phae.: *Dictyopteris delicatula, D. justii, D. plagiogramma, Dictyota dentata, Turbinaria turbinata.* — Rho.: *Amphiroa fragilissima, Centroceras clavulatum, Chondria curvilineata, Gelidiella acerosa, Herposiphonia pecten-veneris, Hypnea* cf. *cervicornis, H. musciformis, H.* cf. *spinella, Jania adhaerens, J. pumila, Laurencia obtusa, L. papillosa, Liagora ceranoides, Liagora* spec., *Melobesia farinosa, Polysiphonia ferulacea.* — Cya.: *Hydrocoleum lyngbyaceum.*

97

1396   GREAT LAGOON, S. of Codrington Village, 4.VII.1955.

Sandy part of large shallow lagoon, with *Batophora, Thalassia, Syringodium* and scattered *Rhizophora*; ½–¾ m.

> Chl.: *Acetabularia crenulata, Anadyomene stellata, Avrainvillea rawsoni, Batophora oerstedi.* — Rho.: *Polysiphonia ferulacea.*

1396A GREAT LAGOON, S. of Codrington Village, 4.VII.1955.

On roots of mangroves in sandy mud; 0–1½ m.

> Chl.: *Acetabularia crenulata, Anadyomene stellata, Batophora oerstedi.* — Rho.: *Jania adhaerens, Laurencia gemmifera, Polysiphonia ferulacea, P.* cf. *subtilissima.*

FOURCHE (W. of St. Barts)

1124   FIVE ISLAND BAY, N. E. shore, 2.VI.1949.

Rocky shore with andesite debris; 0–1½ m.

> Chl.: *Boodlea composita, Chaetomorpha brachygona.* — Phae.: *Dilophus alternans, Padina sanctae-crucis.* — Rho.: *Ceramium byssoideum, Galaxaura rugosa, Jania* spec., *Liagora farinosa.*

ISLOTE AVES (W. of Guadeloupe)

410   EASTERN SHORE, 12.V.1949.

Cast ashore on sandy beach.

> Phae.: *Sargassum fluitans, S. natans.*

1114A NORTHERN LAGOON, 12.V.1949.

Sandy shore with some beachrock; ½–1 m.

> Chl.: *Halimeda opuntia.*

1115   NORTHERN REEF, 12.V.1949.

Beachrock flat; 0–½ m.

> Chl.: *Bryopsis pennata.* — Phae.: *Dictyopteris delicatula, Turbinaria turbinata.* — Rho.: *Centroceras clavulatum, Ceramium byssoideum, Coelothrix irregularis, Gelidiella* spec., *Gelidiopsis planicaulis, Laurencia obtusa, L. papillosa.*

NEVIS

413   FORT CHARLES, S. of Charlestown, 28.VI.1949.

Cast ashore on rocky beach.

> Rho.: *Bryothamnion triquetrum, Hypnea musciformis, Laurencia papillosa, Lophocladia trichoclados, Spyridia aculeata.*

98

1 Near WASHGUT, 27.V.1958.

Andesitic rocks; rather heavy swell; littoral and sublittoral region.

Chl.: *Chaetomorpha media, Cladophora* spec., *Ulva fasciata, Valonia ocellata.* — Phae.: *Dilophus guineensis, Sargassum* spec., *Turbinaria turbinata.* — Rho.: *Ceramium byssoideum, Crouania attenuata, Grateloupia cuneifolia, Hypnea musciformis, Laurencia intricata, L. microcladia, L. papillosa, Laurencia* spec., *Polysiphonia ferulacea, Wrangelia argus.* — Cya.: *Symploca hydnoides.*

2 EAST SHORE, near The Level, 27.V.1958.

Surfswept volcanic rock; sublittoral region.

No algae collected.

3 FLAT POINT, 27.V.1958.

Rockpool; water constantly renewed by heavy swell; littoral region and upper part of sublittoral region (Fig. 20).

Phae.: *Dictyota* spec., *Dilophus guineensis, Padina gymnospora, Sargassum platycarpum, S. polyceratium, Turbinaria turbinata.* — Rho.: *Ceramium byssoideum,* cf. *Dasya* spec., *Laurencia gemmifera, L. microcladia, L.* cf. *obtusa, L. papillosa, Laurencia* spec., *Polysiphonia ferulacea.* — Cya.: *Entophysalis conferta, Lyngbya lutea, L. majuscula, L. rivulariarum, Hydrocoleum glutinosum, Spirulina labyrinthiformis, Symploca hydnoides.*

4 NORTH COAST, near Old Sulphur Mines, 27.V.1958.

Large rock, about 10 m from shore, in rather heavy swell; sublittoral region.

Phae.: *Dictyopteris delicatula, Dictyota jamaicensis.*

5 FORT BAY, from Customs Office westward, 28.V.1958.

Andesitic rock, slightly exposed; sublittoral region.

Phae.: *Colpomenia sinuosa, Dictyopteris delicatula, Dictyota cervicornis, D. jamaicensis, Padina vickersiae.* — Rho.: *Champia parvula, Galaxaura oblongata, G. rugosa, G. subverticillata, Gelidiella acerosa, Grateloupia* spec., *Hypnea cervicornis, Hypnea* spec., *Jania* spec., *Laurencia papillosa, Wrangelia argus.*

6 West of FORT BAY, 28.V.1958.

Andesitic boulders, slightly exposed; littoral region and upper part of sublittoral region (Pl. IXb; Fig. 21).

Chl.: *Chaetomorpha brachygona, C. media, Cladophora* spec., *Diplochaete solitaria.* — Phae.: *Dictyopteris delicatula, Dictyota* cf. *ciliolata, Pocockiella variegata, Sargassum* cf. *vulgare, Turbinaria turbinata.* — Rho.: *Centroceras clavulatum, Ceramium floridanum, Champia parvula, Grateloupia cuneifolia, Kylinia crassipes, Laurencia intricata, L. microcladia, L. papillosa, Laurencia* spec., *Lomentaria rawitscheri, Polysiphonia ferulacea, Spyridia filamentosa, Wrangelia argus.*

99

7 SOUTHWEST COAST, near Parish Hill, 28.V.1958.
Several large boulders in heavy swell; littoral region and upper part of sub-littoral region.

>Chl.: *Chaetomorpha brachygona, C. media.* — Phae.: *Dictyopteris deli-catula.* — Rho.: *Galaxaura subverticillata, Grateloupia filicina, Lopho-siphonia cristata, Laurencia microcladia, L. papillosa, Laurencia* spec. — Cya.: *Hydrocoleum lyngbyaceum, Lyngbya rivulariarum.*

8 SOUTHWEST COAST, near Parish Hill, 28.V.1958.
Volcanic rock; sublittoral region.

>Phae.: *Giffordia mitchellae, Padina vickersiae, Pocockiella variegata.* — Rho.: *Centroceras clavulatum, Galaxaura rugosa, Gelidiella acerosa, Grate-loupia filicina, Jania adhaerens, Liagora decussata, Pterocladia pinnata.*

434 SPRING BAY, 28.VII.1949.
Debris of andestic lavas; cast ashore.

>Chl.: *Chaetomorpha media, Ulva fasciata.* — Rho.: *Centroceras clavulatum, Grateloupia cuneifolia.*

1120 West of FORT BAY, 21.VII.1949.
Rocky coast, andesitic boulders; 0–1½ m (WAGENAAR HUMMELINCK, 1953, pl. VIIIa).

>Phae.: *Dictyopteris delicatula, Dictyota jamaicensis, Giffordia mitchellae, Sargassum* spec., *Turbinaria turbinata.* — Rho.: *Centroceras clavulatum, Ceramium byssoideum, Hypnea* cf. *spinella, Hypnea* spec., *Hypneocolax stellaris, Laurencia obtusa, L. papillosa, Lomentaria rawitscheri, Poly-siphonia ferulacea.*

1120A West of FORT BAY, 21. VII. 1949.
>Andesitic rock; 0–1½ m.
>Rho.: *Hypnea* spec.

ST. BARTS (= Saint-Barthélemy)

1121 South of PUBLIC, near Gustavia, 4.VI.1949.
Rocky shore, andesitic debris with sand; 0–½ m.

>Chl.: *Bryopsis pennata, Caulerpa ambigua, Valonia ventricosa.* — Phae.: *Colpomenia sinuosa, Dictyopteris delicatula, Padina vickersiae, Sargassum* spec. — Rho.: *Amphiroa fragilissima, Gracilaria* cf. *mammillaris, Laurencia microcladia, L. obtusa, L. papillosa, Polysiphonia ferulacea, Wrangelia argus.*

1122 GRANDE SALINE, 3.VI.1949.
Pool in cemented gutter for yearly supply of sea-water; soft mud; turbid water (34 g Cl/l).

>Cya.: *Gomphosphaeria aponina, Lyngbya semiplena.*

ST. CROIX

1404 KRAUSSE LAGOON, seaside of entrance, 15.VI.1955.
Muddy sandflat with scattered *Rhizophora, Thalassia* and *Syringodium*; 0–2 m.

>Chl.: *Batophora oerstedi, Cladophoropsis membranacea, Halimeda opuntia, Penicillus capitatus.* — Rho.: *Melobesia farinosa.*

100

1405 Krausse Lagoon, entrance, 15.VI.1955.

Sand with some mud in narrow canal with tidal flow among *Rhizophora*; 0–1½ m.

Chl.: *Batophora oerstedi, Halimeda incrassata, H. opuntia.* — Rho.: *Acantho-phora spicifera, Centroceras clavulatum, Chondria atropurpurea, Hypnea cervicornis, Melobesia farinosa, Polysiphonia* spec., *Spyridia filamentosa.*

1406 Krausse Lagoon, basin, 15.VI.1955.

Large and shallow mudflat with *Rhizophora, Syringodium, Thalassia* and *Batophora*; 0–1 m.

Chl.: *Batophora oerstedi.*

St. Eustatius

1 Boekaniers Bay, 20.V.1958.

Andesitic rocks; rather heavy swell; littoral region and upper part of sublittoral region.

Chl.: *Chaetomorpha media, Cladophoropsis membranacea.* — Phae.: *Dic-tyota dentata, Dilophus guineensis, Sargassum* spec., *Turbinaria turbinata.* — Rho.: *Centroceras clavulatum, Grateloupia cuneifolia, Hypnea musci-formis, Laurencia microcladia, L. papillosa, Laurencia* spec., *Lophosiphonia* spec., *Polysiphonia ferulacea, Spyridia aculeata.* — Cya.: *Calothrix con-fervicola, C. parietina, Lyngbya semiplena, Symploca muscorum.*

2 Bay between Boekaniers Bay and Corre Corre Bay, 20.V.1958.

Rather heavy swell; littoral region and upper part of sublittoral region.

Chl.: *Chaetomorpha media, Valonia ventricosa.* — Phae.: *Dictyopteris delicatula, Dictyota dentata, D. jamaicensis, Sargassum platycarpum, S.* cf. *vulgare, Stypopodium zonale, Turbinaria turbinata.* — Rho.: *Amphiroa fragilissima, Hypnea* spec.

3 Corre Corre Bay, 20.V.1958.

Quiet water, at landside of coral reef; littoral region and upper part of sub-littoral region.

Chl.: *Caulerpa ambigua, C. racemosa, Cladophoropsis membranacea, Ulva fasciata, Valonia ocellata.* — Phae.: *Dictyopteris delicatula, Dictyota* cf. *cervicornis, Turbinaria turbinata.* — Rho.: *Amphiroa fragilissima, Centro-ceras clavulatum, Ceramium* spec., *Galaxaura cylindrica, G. subverticillata, Gelidiella acerosa, Grateloupia filicina, Jania adhaerens, Laurencia gemmi-fera, L. intricata, L. microcladia, L. obtusa, L. papillosa, Liagora valida, Wrangelia argus.* — Cya.: *Calothrix parietina, Lyngbya lutea, Phormidium papyraceum.*

4 Concordia Bay, 20.V.1958.

Large boulders on sandy beach; rather heavy swell; littoral region and upper part of sublittoral region.

Chl.: *Caulerpa microphysa, Dictyosphaeria vanbosseae, Halimeda tuna, Valonia* cf. *ventricosa.* — Phae.: *Dictyopteris delicatula, D. justii, Dictyota dentata, Dilophus guineensis, Padina* cf. *vickersiae, Sargassum natans,*

Sargassum spec. – Rho.: *Bryothamnion triquetrum, Ceramium byssoideum, Digenea simplex, Dipterosiphonia dendritica, Hypnea musciformis, Jania adhaerens, Laurencia papillosa, Laurencia* spec., *Liagora pinnata, Lophosiphonia cristata, Ochtodes filiformis, Polysiphonia ferulacea, Spermothamnion investiens, Spermothamnion* spec. – Cya.: *Lyngbya majuscula.*

5 BACK-OFF BAY, 21.V.1958.

Boulders along the coast, near Sugarloaf and White Wall; little wave action; littoral region and upper part of sublittoral region (Pl. X).

Chl.: *Chaetomorpha brachygona, C. media, Neomeris annulata, Ulva lactuca.* – Phae.: *Dictyota ciliolata, D. dentata, Dilophus guineensis, Padina* cf. *vickersiae, Sargassum* spec. – Rho.: *Ceramium byssoideum, C. floridanum, Grateloupia cuneifolia, Herposiphonia tenella, Hypnea musciformis, Laurencia microcladia, L. papillosa, Liagora valida, Lophosiphonia cristata, Polysiphonia ferulacea, Spyridia aculeata, Stichothamnion antillarum, Wrangelia argus.* – Cya.: *Lyngbya semiplena, Spirulina labyrinthiformis.*

6 BACK-OFF BAY, 21.V.1958.

Sublittoral region to about 20 m from the coast, large boulders in sandy bottom.

Chl.: *Udotea flabellum.* – Phae.: *Dictyopteris delicatula, Dictyota dentata, Sargassum* spec. – Rho.: *Amphiroa hancockii, Coelothrix irregularis, Galaxaura cylindrica, G. oblongata, G. rugosa, G. subverticillata, Hypnea* spec., *Jania adhaerens, J. pumila, Ochtodes filiformis.*

1116 Southern part of GALLOWS BAY, 15.VII.1949.

Rocky beach, andesite, with pebbles, 0–½ m.

Phae.: *Dictyopteris delicatula, Turbinaria turbinata.* – Rho.: *Gelidiella acerosa.*

1116B Southern part of GALLOWS BAY, 15.VII.1949.

Rocky beach; 1–2 m.

Chl.: *Cladophoropsis membranacea, Halimeda tuna, Struvea anastomosans, Udotea flabellum, Valonia ventricosa.* – Phae.: *Dictyopteris delicatula, Dictyota* spec., *Dilophus alternans, Pocockiella variegata, Sargassum platycarpum, S. polyceratium, Turbinaria turbinata.* – Rho.: *Amphiroa fragilissima, A. hancockii, Coelothrix irregularis, Digenea simplex, Galaxaura marginata, G. oblongata, G. squalida, G. subverticillata, Gelidiella acerosa, Gracilaria* spec., *Hypnea* cf. *cervicornis, Jania* spec., *Liagora farinosa, Wrangelia argus.*

1117 DOWNTOWN, near Billy Gut, 13.VII.1949.

Sandy shore, on andesite rock; 0–½ m.

Chl.: *Caulerpa sertularioides, Chaetomorpha brachygona, Enteromorpha lingulata.* – Phae.: *Dictyota ciliolata.* – Rho.: *Callithamnion byssoides, Ceramium fastigiatum, Champia parvula, Galaxaura* cf. *rugosa, Grateloupia filicina, Herposiphonia tenella, Jania adhaerens, J. rubens, Polysiphonia ferulacea, Wurdemannia miniata.*

1118 BILLY GUT, near Downtown, 13.VII.1949.

Sandy shore, on andesite rock; ½–1½ m.

Chl.: *Chaetomorpha clavata, C. media, Enteromorpha flexuosa.*

102

1119 South of Tumble Down Dick Bay, 10.VII.1949.

Rocky andesite shore with magnetite sand; 0–½ m.

> Chl.: *Chaetomorpha media.* – Phae.: *Dictyota ciliolata, Dictyopteris delicatula, Dilophus guineensis, Giffordia mitchellae, Sargassum polyceratium.* – Rho.: *Gracilaria mammillaris, Herposiphonia tenella, Hypnea musciformis, Jania adhaerens, Laurencia microcladia, L. papillosa, Laurencia spec., Lophosiphonia cristata, Polysiphonia ferulacea, Pterocladia bartlettii, Spyridia aculeata.*

St. John

1407 Turner Bay, E. part, 18.VI.1955.

Surfswept porphyritic rock, boulders and coarse sand; 0–½ m.

> Chl.: *Cladophoropsis membranacea, Halimeda opuntia, Valonia ventricosa.* – Phae.: *Dictyota cervicornis, D. jamaicensis, Padina* cf. *gymnospora, P. sanctae-crucis, Pocockiella variegata, Turbinaria turbinata.* – Rho.: *Amphiroa fragilissima, Coelothrix irregularis, Crouania attenuata, Jania adhaerens, Laurencia microcladia, L. papillosa, Liagora ceranoides.* – Cya.: *Dichothrix fucicola.*

1408 Bay south of Cruz Bay, 19.VI.1955.

Boulders on sandy beach; ½–1½ m.

> Chl.: *Cladophoropsis membranacea.* – Phae.: *Dictyota ciliolata, Padina sanctae-crucis, Sargassum polyceratium, Sargassum spec., Turbinaria turbinata.* – Rho.: *Centroceras clavulatum, Galaxaura subverticillata, Herposiphonia tenella, Jania adhaerens, Laurencia microcladia, L. obtusa, Lophosiphonia cristata, Melobesia farinosa, Polysiphonia ferulacea, P. howei.* – Cya.: *Dichothrix fucicola.*

St. Kitts (= St. Christopher)

1397 Frigate Bay, S. E. part, 20.VII.1959.

Exposed cliff of volcanic tuff, boulders and coarse sand; 0–½ m.

> Chl.: *Caulerpa ambigua, C. sertularioides, Cladophoropsis membranacea, Ernodesmis verticillata, Valonia ventricosa.* – Phae.: *Dictyota ciliolata, D. divaricata, Dictyota spec., Dilophus guineensis, Padina sanctae-crucis.* – Rho.: *Amphiroa fragilissima, Centroceras clavulatum, Gelidiopsis intricata, Hypnea musciformis, H. spinella, Laurencia papillosa, Liagora* cf. *ceranoides, L.* cf. *pedicellata, Melobesia farinosa, Polysiphonia ferulacea, P. howei, Wrangelia argus.*

1398 West of Basseterre, near St. Thomas Point, 30.VI.1955.

Cliff of andesitic rock, pebbles and debris; 0–½ m.

> Chl.: *Bryopsis plumosa, Chaetomorpha media.* – Rho.: *Champia parvula, Gelidium pusillum, Hypnea spinella, Wrangelia argus.*

St. Martin

1 Little Bay, eastern slope of Kay Bay Hill, 28.IV.1958.

Boulders of Point Blanche formation; rather heavy swell; littoral region and upper part of sublittoral region (Pl. VIIa; Fig. 14).

Chl.: *Anadyomene stellata, Caulerpa ambigua, Chaetomorpha media, Cladophora luteola, Cladophoropsis membranacea, Dictyosphaeria vanbosseae, Halimeda opuntia, Struvea anastomosans, Valonia ocellata, V. ventricosa.* — Phae.: *Dictyopteris delicatula, Dictyota dentata, Dilophus guineensis, Ectocarpus breviarticulatus, Giffordia duchassaingiana, Padina sanctae-crucis, Pocockiella variegata, Sargassum* cf. *filipendula, Sargassum* spec., *Turbinaria turbinata.* — Rho.: *Amphiroa fragilissima, Asparagopsis taxiformis, Centroceras clavulatum, Ceramium floridanum, Dasya* cf. *rigidula, Falkenbergia hillebrandii, Herposiphonia pecten-veneris, H. tenella, Hypnea musciformis, Jania adhaerens, Laurencia gemmifera, L. microcladia, L. papillosa, Melobesia farinosa, Ochtodes filiformis, Polysiphonia ferulacea, Spyridia aculeata, Wrangelia argus.* — Cya.: *Dichothrix fucicola, Symploca hydnoides.*

2 LITTLE BAY, E. side, on peninsula with Fort Amsterdam, 29.IV.1958.

Dolerite rocks at waterlevel; slightly exposed; littoral region and upper part of sublittoral region (Pl. IIIb; Fig. 14).

Chl.: *Boodlea composita, Chaetomorpha media, Cladophoropsis membranacea, Dictyosphaeria vanbosseae, Neomeris annulata.* — Phae.: *Dictyota bartayresii, D. ciliolata, D. divaricata, Dilophus guineensis, Ectocarpus breviarticulatus, E. confervoides, Padina sanctae-crucis, P. vickersiae, Pocockiella variegata, Sargassum polyceratium, Turbinaria turbinata.* — Rho.: *Acanthophora spicifera, Centroceras clavulatum, Crouania attenuata, Falkenbergia hillebrandii, Gymnogongrus tenuis, Jania adhaerens, Laurencia microcladia, L. obtusa, L. papillosa, Peyssonnelia* spec., *Polysiphonia ferulacea, P. howei, Polysiphonia* spec., *Spyridia filamentosa, Wrangelia argus.* — Cya.: *Coccochloris stagnina, Lyngbya lutea, L. semiplena, Scytonema myochrous, Spirulina subsalsa.*

3 LITTLE BAY, 29.IV.1958.

Boulders in front of dolerite coast; slightly exposed; sublittoral region (Pl. IIIb; Fig. 14).

Phae.: *Dilophus alternans, Turbinaria turbinata.* — Rho.: *Dasya* spec., *Galaxaura* spec., *Laurencia obtusa, Wrangelia argus.*

4 LITTLE BAY, E. side, 30.IV.1958.

Boulders of coral limestone constantly washed by the waves; much whirling sand; littoral region and upper part of sublittoral region (Fig. 14).

Chl.: *Acetabularia calyculus, Caulerpa sertularioides, Chaetomorpha clavata, Cladophora fuliginosa, Cladophoropsis membranacea, Neomeris annulata, Penicillus capitatus, P. lamourouxii, Valonia ventricosa.* — Phae.: *Dictyota dichotoma, Dilophus guineensis, Padina sanctae-crusis, P. vickersiae, Sargassum* cf. *filipendula.* — Rho.: *Ceramium byssoideum, Chondria curvilineata, C. tenuissima, Dasya* cf. *corymbifera, Galaxaura squalida, Laurencia* cf. *microcladia, L. papillosa, L. poitei, Liagora farinosa, Lophosiphonia cristata, Polysiphonia ferulacea.* — Cya.: *Hydrocoleum lyngbyaceum, Lyngbya rivulariarum, Spirulina labyrinthiformis.*

5 LITTLE BAY, E. of beach, 30.IV.1958.

Coral limestone; slightly exposed; littoral region and upper part of sublittoral region (Pl. IIIb; Fig. 14).

Chl.: *Anadyomene stellata, Bryopsis pennata, Caulerpa ambigua, Clado-phora luteola, Cladophoropsis membranacea, Halimeda opuntia.* – Phae.: *Dictyopteris delicatula, Dictyota ciliolata, Dilophus guineensis, Padina sanctae-crucis, Sargassum vulgare.* – Rho.: *Asparagopsis taxiformis, Bryo-thamnion triquetrum, Chondria tenuissima, Falkenbergia hillebrandii, Griffithsia tenuis, Heterosiphonia wurdemanni, Laurencia obtusa, L.* cf. *obtusa, L. papillosa, Laurencia* spec., *Lophosiphonia cristata, Wrangelia argus.* – Cya.: *Dichothrix fucicola.*

6 LITTLE BAY, 30.IV.1958.

Rocks along the beach; sublittoral region (Fig. 14).

Chl.: *Anadyomene stellata, Caulerpa sertularioides, Cladophora fuliginosa, Dictyosphaeria vanbosseae, Halimeda incrassata, Neomeris annulata, Peni-cillus capitatus, Valonia ventricosa.* – Phae.: *Padina sanctae-crucis, Sar-gassum* spec., – Rho.: *Amphiroa fragilissima, Dasya rigidula, Dasya* spec., *Digenea simplex, Gelidiella acerosa, Jania adhaerens, Laurencia papillosa, Polysiphonia* spec.

7 GREAT BAY, E. side of peninsula with Fort Amsterdam, 2.V.1958.

Steep cliff of dolerite rock; heavy swell; littoral region and upper part of sublittoral region (Pl. IIIa; Fig. 14 & 17).

Chl.: *Caulerpa ambigua, C. racemosa, C. sertularioides, Chaetomorpha media, Cladophoropsis membranacea, Dictyosphaeria vanbosseae, Diplo-chaete solitaria, Halimeda opuntia.* – Phae.: *Dictyota cervicornis, D. jamaicensis, Dilophus guineensis, Ectocarpus breviarticulatus, E. confer-voides, Giffordia rallsiae, Padina sanctae-crucis, Pocockiella variegata, Sargassum vulgare.* – Rho.: *Acanthophora spicifera, Bostrychia binderi, Centroceras clavulatum, Gelidiella acerosa, Gracilaria mammillaris, Grate-loupia cuneifolia, G. filicina, Hypnea musciformis, Jania adhaerens, Lauren-cia microcladia, L. obtusa, L.* cf. *obtusa, L. papillosa, L. scoparia, Ochtodes filiformis, Polysiphonia ferulacea, P. howei, Pterocladia pinnata, Wrangelia argus.* – Cya.: *Dermocarpa prasina, Symploca hydnoides.*

8 SIMSON BAY, E. part, 6.V.1958.

Rocks of Low Lands formation and of Point Blanche formation on sandy bottom with seagrass and algae; slightly exposed; littoral region and upper part of sublittoral region (Pl. Vb; Fig. 16).

Chl.: *Acetabularia crenulata, A. polyphysoides, Anadyomene stellata, Avrainvillea rawsoni, Caulerpa cupressoides, C. racemosa C. sertularioides, Chaetomorpha brachygona, Cladophora fuliginosa, C. uncinata, Cladopho-ropsis membranacea, Dictyosphaeria cavernosa, Halimeda discoidea, H. incrassata, H. opuntia, Neomeris annulata, Penicillus capitatus, P. dumetosus, Rhipocephalus phoenix, Udotea flabellum.* – Phae.: *Dictyopteris delicatula, Dictyota ciliolata, D. dentata, D. dichotoma, D. divaricata, Dilophus alter-nans, D. guineensis, Giffordia rallsiae, Padina sanctae-crucis, Sargassum platycarpum, Stypopodium zonale, Turbinaria turbinata.* – Rho.: *Acantho-phora spicifera, Amphiroa fragilissima, A. rigida* var. *antillana, Bryotham-nion triquetrum, Chondria tenuissima, Crouania pleonospora, Digenea simplex, Galaxaura rugosa, G. subverticillata, Gracilaria debilis, G. ferox, Herposiphonia secunda, H.* cf. *tenella, Jania pumila, Laurencia gemmifera, L. microcladia, L. obtusa, L. papillosa, Laurencia* spec., *Liagora pedicel-lata, L. valida, Ochtodes filiformis, Polysiphonia* spec.

9   Simson Bay, E. part, 6.V.1958.

Seagrass and algae vegetation; sublittoral region (Fig. 16).

Chl.: *Caulerpa sertularioides, Cladophoropsis membranacea, Dictyosphaeria cavernosa, Halimeda discoidea, H. incrassata, Penicillus capitatus, P. lamourouxii, Udotea flabellum.* — Phae.: *Dictyota dichotoma, Dictyota* spec., *Dilophus guineensis, Padina sanctae-crucis.* — Rho.: *Acanthophora spicifera, Chondria curvilineata, Crouania pleonospora, Hypnea* cf. *cervicornis, Laurencia obtusa, Melobesia farinosa, Polysiphonia binneyi, P.* cf. *gorgoniae.*

10   Burgeux Bay, W. part, 7.V.1958.

Rocks of Low Lands formation and beachrock; heavy swell; littoral region and upper part of sublittoral region (Pl. VIa; Fig. 16).

Chl.: *Anadyomene stellata, Caulerpa sertularioides, Chaetomorpha media, Cladophora fuliginosa, Cladophoropsis membranacea, Codium isthmocladum, Dictyosphaeria cavernosa, D. vanbosseae, Halimeda discoidea, H. opuntia, Neomeris annulata, Rhipocephalus phoenix.* — Phae.: *Colpomenia sinuosa, Dictyopteris delicatula, Dictyota ciliolata, D. dentata, Dilophus alternans, D. guineensis, Ectocarpus elachistaeformis, Giffordia mitchellae, Padina sanctae-crucis, Sargassum platycarpum, S. rigidulum, S. vulgare, Stypopodium zonale, Turbinaria turbinata.* — Rho.: *Acanthophora spicifera, Centroceras clavulatum, Ceramium cruciatum, Chondria* cf. *tenuissima, Corallina cubensis, Galaxaura rugosa, Herposiphonia* cf. *tenella, Hypnea musciformis, Jania adhaerens, Laurencia microcladia, L. obtusa, L. papillosa, Laurencia* spec., *Lophosiphonia cristata, Lophosiphonia* spec., *Polysiphonia ferulacea, P. howei, Spyridia aculeata, S. filamentosa, Wrangelia argus.* — Cya.: *Hormothamnion enteromorphoides, Hydrocoleum lyngbyaceum, Lyngbya rivulariarum, Rivularia polyotis, Spirulina labyrinthiformis.*

11   Cole Bay, 8.V.1958.

Sandy beach with beachrock; heavy swell; littoral region and upper part of sublittoral region (Pl. Va).

Chl.: *Cladophora fuliginosa, C. uncinata, Dictyosphaeria cavernosa.* — Phae.: *Dictyota dentata, Dilophus guineensis, Padina sanctae-crucis, Sargassum* spec. — Rho.: *Bryothamnion triquetrum, Chondria* cf. *tenuissima, Dasya rigidula, Gracilaria debilis, Laurencia* spec., *Lophosiphonia cristata, Polysiphonia ferulacea.*

12   Maho Bay, 9.V.1958.

Flat rock of Low Lands formation, about 50 cm above sea-level, washed by heavy swell; littoral region and upper part of sublittoral region (Pl. VIb; Fig. 16).

Chl.: *Anadyomene stellata, Cladophora fuliginosa, Cladophoropsis membranacea.* — Phae.: *Dilophus guineensis, Padina sanctae-crucis, Sargassum* cf. *filipendula, Turbinaria turbinata.* — Rho.: *Bryothamnion triquetrum, Chondria tenuissima, Dipterosiphonia dendritica, Herposiphonia secunda, Laurencia microcladia, L. papillosa, Ochtodes filiformis, Polysiphonia ferulacea, Spyridia aculeata, Wrangelia argus.* — Cya.: *Dichothrix fucicola, Hydrocoleum lyngbyaceum, Lyngbya rivulariarum, Phormidium valderianum, Spirulina labyrinthiformis.*

13 MAHO BAY, 9.V.1958.

Steep cliff of Low Lands formation; a flat rock at about 3–4 m; sublittoral region (Fig. 16).

Chl.: *Halimeda opuntia, Rhipocephalus phoenix, Udotea flabellum, U. sublittoralis.* — Phae.: *Dilophus alternans.* — Rho.: *Amphiroa hancockii, Ceramium byssoideum, Champia parvula, Galaxaura rugosa, Griffithsia tenuis, Jania adhaerens, Laurencia obtusa, Ochtodes filiformis, Spermothamnion investiens, Wrangelia argus.*

14 OYSTER POND, bay, 12.V.1958.

Boulders of Point Blanche formation; heavy swell; littoral region and upper part of sublittoral region (Pl. VIIIa; Fig. 15).

Chl.: *Caulerpa ambigua, Chaetomorpha media, Cladophoropsis membranacea, Enteromorpha flexuosa, Halimeda opuntia.* — Phae.: *Dictyopteris delicatula, Dictyota* cf. *ciliolata, Giffordia mitchellae, Dilophus guineensis, Sargassum platycarpum, S. polyceratium, Sargassum* spec., *Turbinaria turbinata.* — Rho.: *Acrochaetium seriatum, Bryothamnion triquetrum, Centroceras clavulatum, Chondria tenuissima, Corallina cubensis, Digenea simplex, Gracilaria ferox, G.* cf. *mammillaris, Grateloupia filicina, Herposiphonia tenella, Heterosiphonia wurdemanni, Hypnea musciformis, Jania rubens, Laurencia* cf. *obtusa, L. papillosa, Laurencia* spec., *Lophosiphonia cristata, Polysiphonia ferulacea, Wrangelia argus.* — Cya.: *Lyngbya majuscula.*

15 OYSTER POND, lagoon, 12.V.1958.

Coast of coral limestone; slight wave action; littoral region and upper part of sublittoral region to a depth of 40 cm; partly a mangrove vegetation with *Thalassia* in sandy bottom (Fig. 15).

Chl.: *Avrainvillea rawsoni,* cf. *Boodleopsis pusilla, Caulerpa sertularioides, Chaetomorpha crassa, C. linum, C. media, Cladophora fascicularis, Cladophoropsis membranacea, Codium taylori, Dictyosphaeria vanbosseae, Ernodesmis verticillata, Enteromorpha compressa, E. lingulata, Halimeda opuntia, Penicillus capitatus, Struvea anastomosans, Ulva lactuca.* — Phae.: *Dictyopteris delicatula, Dictyota* spec., *Padina gymnospora, Sargassum polyceratium.* — Rho.: *Acanthophora spicifera, Amphiroa fragilissima, Bostrychia binderi, B. moritziana, Caloglossa leprieurii, Centroceras clavulatum, Galaxaura squalida, Gracilaria cervicornis, G. mammillaris, Hypnea cervicornis, Laurencia papillosa, Murrayella periclados, Polysiphonia howei, Pterocladia pinnata, Spyridia filamentosa.* — Cya.: *Entophysalis conferta, Lyngbya majuscula.*

16 OYSTER POND, lagoon, 12.V.1958.

Detached algae, floating in the lagoon, probably coming from outer bay (Fig. 15).

Chl.: *Chaetomorpha crassa, Cladophora fascicularis.* — Phae.: *Dictyopteris delicatula, Dictyota* spec., *Dilophus alternans, Pocockiella variegata.* — Rho.: *Bryothamnion seaforthii, Centroceras clavulatum, Ceramium fastigiatum, Galaxaura marginata, Lomentaria* spec.

17 GUANA BAY, 13.V.1958.

Coast of coral limestone; heavy swell; littoral region and upper part of sublittoral region (Pl. IVa and VIIIb; Fig. 11).

107

Chl.: *Avrainvillea rawsoni, Cladophora luteola, Cladophoropsis membranacea, Halimeda opuntia, Rhipocephalus phoenix,* cf. *Valonia ocellata.* — Phae.: *Dictyopteris plagiogramma, Dictyota dentata, Dilophus guineensis, Padina sanctae-crucis, Sargassum platycarpum, Stypopodium zonale, Turbinaria turbinata.* Rho.: *Centroceras clavulatum, Chondria tenuissima, Digenea simplex, Hypnea musciformis, Herposiphonia tenella, Jania adhaerens, Laurencia microcladia, L. papillosa, Lophocladia trichoclados, Lophosiphonia cristata, Polysiphonia ferulacea, P.* cf. *havanensis, P.* cf. *howei, Thuretia borneti.* — Cya.: *Hydrocoleum lyngbyaceum, Spirulina subsalsa.*

18 GUANA BAY, 13.V.1958.

Rocks of Point Blanche formation, constantly washed by the waves; littoral region and upper part of sublittoral region (Pl. VIIIb; Fig. 10).

Chl.: *Avrainvillea rawsoni, Caulerpa cupressoides, Cladophoropsis membranacea, Dictyosphaeria vanbosseae, Halimeda opuntia, Udotea conglutinata, Valonia aegagropila.* — Phae.: *Dictyopteris delicatula, Dictyota dentata, Dilophus alternans, D. guineensis, Padina sanctae-crucis, Sargassum platycarpum, Sargassum* spec., *Turbinaria turbinata.* — Rho.: *Amphiroa fragilissima, Herposiphonia tenella, Hypnea musciformis, Laurencia microcladia, L. papillosa, Lophosiphonia cristata, Polysiphonia ferulacea, Polysiphonia* spec. — Cya.: *Hydrocoleum lyngbyaceum, Lyngbya rivulariarum, Symploca hydnoides.*

19 GUANA BAY, 13.V.1958.

Detached algae, floating near sandy beach, S. of rock of coral limestone.

Chl.: *Caulerpa prolifera, Codium isthmocladum.* — Phae.: *Dictyopteris plagiogramma, Dictyota dentata, Dilophus alternans, Sargassum natans, Stypopodium zonale.* — Rho.: *Asparagopsis taxiformis, Bryothamnion triquetrum, Callithamnion cordatum, Chondria tenuissima, Dasya rigidula, D. sertularioides, Enantiocladia duperreyi, Herposiphonia tenella, Heterosiphonia gibbesii, H. wurdemannii, Jania adhaerens, J. pumila, Laurencia obtusa, L.* cf. *obtusa, Lophocladia trichoclados, Polysiphonia ferulacea.*

20 POINT BLANCHE BAY, 14.V.1958.

Detached algae (Fig. 12).

Phae.: *Pocockiella variegata.*

21 GREAT BAY, landing stage near Point Blanche, 14.V.1958.

Boulders of Point Blanche formation; slightly exposed; littoral region and upper part of sublittoral region (Pl. VIIb; Fig. 12).

Chl.: *Cladophora fuliginosa, Cladophoropsis membranacea, Halimeda opuntia, Neomeris annulata.* — Phae.: *Dictyota dentata, Dilophus alternans, D. guineensis, Padina sanctae-crucis, Sargassum platycarpum, S. vulgare, Turbinaria turbinata.* — Rho.: *Galaxaura squalida.* — Cya.: *Dichothrix fucicola, Hydrocoleum lyngbyaceum, Lyngbya rivulariarum.*

22 CUPECOY BAY, 16.V.1958.

Beachrock; heavy swell, much sand scouring the rocks; littoral region and upper part of sublittoral region (Fig. 16).

Chl.: *Avrainvillea rawsoni, Caulerpa ambigua, C. sertulariodes, Cladophora*

fuliginosa, C. uncinata, Cladophoropsis membranacea, Halimeda opuntia.
– Phae.: *Dilophus guineensis, Ectocarpus confervoides, Padina sanctae-crucis, Sargassum platycarpum, Sargassum* spec. – Rho.: *Bryothamnion triquetrum, Chondria curvilineata, C. tenuissima, Lophosiphonia cristata, Polysiphonia ferulacea, Spyridia aculeata.* – Cya.: *Dichothrix fucicola, Hydrocoleum lyngbyaceum, Lyngbya rivulariarum.*

23 CUPECOY BAY, W. part, 16.V.1958.

Rocks of Low Lands formation forming a terrace 1–2 m wide; heavy swell; littoral region and upper part of sublittoral region (Fig. 16).

Phae.: *Dilophus guineensis, Padina sanctae-crucis, P. vickersiae.* – Rho.: *Chondria tenuissima, Jania adhaerens, Herposiphonia tenella, Laurencia microcladia, L. papillosa, Lophosiphonia cristata, Polysiphonia ferulacea, Spyridia aculeata.* – Cya.: *Dichothrix fucicola.*

24 GREAT BAY, W. shore near Little Bay Hotel, 22.V.1958.

Dolerite rocks, partly on sandy bottom; rather heavy swell; sublittoral region (Fig. 14).

Chl.: *Bryopsis pennata, Caulerpa prolifera, C. racemosa, C. sertularioides, C. taxifolia, Chaetomorpha media, Cladophora* spec., *Cladophoropsis membranacea, Ernodesmis verticillata, Halimeda discoidea, H. incrassata, H. opuntia, Penicillus capitatus, P. dumetosus, Struvea anastomosans, Udotea conglutinata, U. flabellum, Valonia ventricosa.* – Phae.: *Dictyota divaricata.* – Rho.: *Amphiroa fragilissima, Centroceras clavulatum, Ceramium byssoideum, Champia parvula, Crouania attenuata, Gracilaria mammillaris, Hypnea spinella, Jania adhaerens, Laurencia obtusa, Melobesia farinosa, Pterocladia pinnata, Wrangelia argus.* – Cya.: *Lyngbya lutea, L. majuscula.*

25 GREAT BAY, W. shore near annex of Little Bay Hotel, 22.V.1958.

Dolerite rocks; rather heavy swell; littoral region and upper part of sublittoral region (Pl. IIIa and IIIb; Fig. 14).

Chl.: *Chaetomorpha media, Struvea anastomosans, Ulva* cf. *lactuca.* – Phae.: *Dictyota jamaicensis, Ectocarpus confervoides, Giffordia rallsiae.* – Rho.: *Centroceras clavulatum, Hypnea musciformis, Laurencia intricata, L. obtusa.*

26 Outlet of FISH POND in Mouth Piece Bay, 23.V.1958.

Rhizophora vegetation; watertemperature rather high; littoral and sublittoral region (Fig. 15).

Chl.: *Chaetomorpha brachygona, Diplochaete solitaria, Enteromorpha chaetomorphoides, Halimeda opuntia, Udotea flabellum.* – Rho.: *Acanthophora spicifera, Asterocytis ramosa, Centroceras clavulatum, Chondria curvilineata, C.* cf. *tenuissima, Kylinia crassipes, Melobesia farinosa, Polysiphonia denudata, Spyridia filamentosa.*

27 MOUTH PIECE BAY (= Anse de l'Embouchure), near outlet of Fish Pond, 23.V.1958.

Rocks in littoral region (Fig. 15).

Chl.: *Cladophora fascicularis, Cladophora* spec., *Enteromorpha erecta.* – Rho.: *Acanthophora spicifera, Centroceras clavulatum, Chondria collinsiana,*

C. curvilineata, C. cf. *tenuissima, Gracilaria verrucosa, Hypnea cornuta, Laurencia papillosa, Spyridia filamentosa.* — Cya.: *Symploca hydnoides.*

28   MOUTH PIECE BAY (= Anse de l'Embouchure), 23.V.1958.

Sandy bottom with seagrasses and algae; *Thalassia* forming banks to about 50 cm above surrounding bottom; sublittoral region (Fig. 15).

Chl.: *Caulerpa crassifolia, C. cupressoides, C. sertularioides, Chaetomorpha aerea, Cladophoropsis membranacea, Dictyosphaeria cavernosa, Ernodesmis verticillata, Halimeda incrassata, H. opuntia, Penicillus capitatus, Udotea flabellum, Valonia aegagropila, V. macrophysa, V. ocellata, V. ventricosa.* — Phae.: *Dictyota divaricata, Dictyota* spec. — Rho.: *Amphiroa fragilissima, Chondria tenuissima, Corallina cubensis, Gracilaria* spec., *Melobesia farinosa.* — Cya.: *Lyngbya majuscula.*

29   SIMSON BAY LAGOON, FLAMINGO POND, 24.V.1958.

Muddy bottom on rocks of Low Lands formation with *Rhizophora* and *Avicennia*; scanty algal growth (Fig. 16).

Chl.: *Batophora oerstedi.*

29a   SIMSON BAY LAGOON, MULLET POND, 24.V.1958.

Muddy bottom with mangroves (Fig. 16).

Chl.: *Batophora oerstedi* (not collected).

29b   SIMSON BAY LAGOON, S.W. part, near Mary Point, 16.V.1958.

Muddy bottom with *Avicennia*; shore vegetation with much *Batis* (Fig. 16).

No algae present.

29c & d   OLD BUILDING POND, S. part, 24.V.1958.

No algae present.

30   SIMSON BAY LAGOON, near building of Airport, 24.V.1958.

Sandy bottom with *Rhizophora* and some *Avicennia* (Fig. 16).

Chl.: *Batophora oerstedi, Enteromorpha compressa, Rhizoclonium kerneri.*

31   SIMSON BAY LAGOON, near former outlet, 24.V.1958.

Narrow creeks on sandflats with *Salicornia, Sesuvium* and *Conocarpus*; 10—20 cm (Fig. 16).

Chl.: *Enteromorpha intestinalis.*

32   BAIE DE LA GRANDE CASE, 26.V.1958.

Beachrock; slightly exposed; littoral region and upper part of sublittoral region (Pl. IXa; Figs. 18 and 19).

A rich algal vegetation (material lost).

33   BAIE DE LA GRANDE CASE, 26.V.1958.

Dioritic rock; slight wave action; littoral region and upper part of sublittoral region (Pl. IXa; Fig. 18).

Chl.: *Chaetomorpha media, Cladophoropsis membranacea, Halimeda*

110

opuntia, Valonia ocellata. — Phae.: *Dictyota dentata, Dilophus guineensis, Padina sanctae-crucis, Pocockiella variegata, Sargassum polyceratium, Turbinaria turbinata.* — Rho.: *Coelothrix irregularis, Digenea simplex, Jania adhaerens, Laurencia gemmifera, L. microcladia, L. obtusa, L. papillosa, Lophosiphonia* spec. — Cya.: *Hydrocoleum glutinosum, Lyngbya rivulariarum.*

34 BAIE DE LA GRANDE CASE, 26.V.1958.

Dioritic rock, sublittoral region (Fig. 18).

Chl.: *Caulerpa sertularioides, Cladophoropsis membranacea, Dictyosphaeria vanbosseae, Halimeda discoidea, H. incrassata, H. opuntia, Neomeris annulata, N. mucosa, Penicillus lamourouxii, P. pyriformis, Rhipocephalus phoenix, Udotea flabellum, Valonia ventricosa.* — Phae.: *Dictyota bartayresii, Dilophus alternans, D. guineensis, Padina sanctae-crucis, Sargassum platycarpum.* — Rho.: *Amphiroa fragilissima, Ceramium nitens, Champia parvula, Chondria curvilineata, Coelothrix irregularis, Dasya pedicellata, Digenea simplex, Galaxaura rugosa, G. subverticillata, Gelidiella acerosa, Jania adhaerens, Laurencia gemmifera, L. obtusa, L. papillosa, Liagora ceranoides, L. farinosa, L. pinnata, L. valida, Liagora* spec., *Lophosiphonia* spec., *Polysiphonia binneyi.* — Cya.: *Lyngbya lutea.*

35 BAIE DE LA GRANDE CASE, 26.V.1958.

Sandy bottom; *Thalassia* vegetation, about 2 m (Fig. 18).

Chl.: *Chaetomorpha linum, Halimeda incrassata, Penicillus capitatus, Struvea anastomosans, Udotea flabellum.* — Phae.: *Dictyota cervicornis, D. indica, Dilophus guineensis, Padina sanctae-crucis.* — Rho.: *Ceramium floridanum, C. nitens, Chondria* cf. *atropurpurea, Dasya pedicellata, D. rigidula, Laurencia gemmifera, L. obtusa, L. papillosa, Liagora valida, Lophosiphonia* spec., *Polysiphonia* cf. *ferulacea.* — Cya.: *Dichothrix fucicola, D. penicillata, Lyngbya lutea.*

542 DEVIL'S HOLE SWAMP, S.E. of Simson Bay Bridge, 4.VIII.1949.

Stagnant pool with tidal movements in sink hole of at least 40 x 20 m, 150—200 m from the shore; limestone and mud, *Avicennia;* water turbid, greenish brown (13.8 g Cl/l).

Chl.: *Batophora oerstedi, Chaetomorpha gracilis.*

1125 GREAT BAY, near Point Blanche, 26.VI. 1949.

Rocky shore, tuffs and tertiary limestone; tide pools; 0—½ m.

Chl.: *Cladophora fuliginosa, Cladophoropsis membranacea, Dictyosphaeria cavernosa.* — Phae.: *Dictyota dentata, Dilophus alternans, D. guineensis, Padina sanctae-crucis, Pocockiella variegata, Sargassum platycarpum, Sargassum* spec., *Sphacelaria tribuloides, Turbinaria turbinata.* — Rho.: *Gelidiella acerosa, Herposiphonia tenella, Jania adhaerens, Laurencia papillosa.*

1125A GREAT BAY, near Point Blanche, 26.VI.1949.

Rocky shore, tuffs and tertiary limestone; rock-pools; ¹/₂—1½ m.

Chl.: *Cladophoropsis membranacea, Dictyosphaeria cavernosa, Valonia ventricosa.* — Cya.: *Dichothrix fucicola.*

111

1126   GREAT BAY, E. shore, 11.VI.1949.

Rocky beach, with some muddy sand; few *Thalassia*; tide pools; 0–1 m (WAGENAAR HUMMELINCK, 1953, pl. VIIIb).

Chl.: *Bryopsis* spec., *Caulerpa racemosa, Cladophoropsis membranacea, Dictyosphaeria cavernosa, Halimeda incrassata, H. opuntia, Valonia aegagropila, V. ocellata, V. ventricosa.* – Phae.: *Dictyota bartayresii.* – Rho.: *Amphiroa fragilissima, Laurencia obtusa, L. papillosa, Melobesia farinosa.*

1127   GREAT BAY, N.E. shore, 16.V.1949.

Rocky beach, debris with muddy sand; *Thalassia*; ½–1½ m.

Chl.: *Caulerpa cupressoides, C. sertularioides, Halimeda incrassata, H. opuntia, Penicillus capitatus, Udotea flabellum.* – Phae.: *Dictyota bartayresii, D. cervicornis, D. divaricata, Sargassum polyceratium.* – Rho.: *Amphiroa fragilissima, Gracilaria* spec., *Hypnea musciformis, Laurencia papillosa, Melobesia farinosa.*

1128A   GREAT BAY, Philipsburg, 26.V.1949.

Wooden wreck on sand beach; 0–1½ m.

Phae.: *Dictyopteris delicatula, Dictyota cervicornis, D. indica.* – Rho.: *Acanthophora spicifera, Gracilaria cervicornis, Hypnea* cf. *cervicornis, Hypnea musciformis.*

1128Aa   GREAT BAY, Philipsburg, 24.VI.1955.

Wooden wreck on sand beach; 0–1½ m.

Chl.: *Cladophoropsis membranacea, Ernodesmis verticillata, Halimeda* cf. *opuntia.* – Phae.: *Dictyota* spec., *Padina vickersiae.* – Rho.: *Acanthophora spicifera, Dasya rigidula, Gracilaria* cf. *domingensis, G. ferox, Hypnea cervicornis, H. musciformis, Jania adhaerens. Polysiphonia howei.*

1128B   GREAT BAY, Philipsburg, 26.VI.1949.

Detached *Ulva* and other weeds on sand beach; ¾–1½ m.

Chl.: *Ulva lactua.* – Phae.: *Dictyota* spec. – Rho.: *Hypnea musciformis, Hypneocolax stellaris.*

1128C   GREAT BAY, N.E. shore, 14.VI.1949.

Sand beach with *Thalassia*; 1½–2½ m.

Chl.: *Blastophysa rhizopus, Ulva* cf. *lactuca.* – Rho.: *Herposiphonia secunda, Hypnea musciformis, Melobesia farinosa.*

1129   SIMSON BAY BRIDGE, 4.VIII.1949.

On wooden piles in sand of lagoon entrance, *Thalassia*; strong tidal flow; 0–1½ m.

Chl.: *Caulerpa racemosa, C. sertularioides, Halimeda opuntia.* – Phae.: *Dictyota* cf. *bartayresii.* – Rho.: *Hypnea cervicornis, Jania adhaerens, Laurencia gemmifera.*

1130 SIMSON BAY LAGOON, outlet, 27.V.1949.

Sandy lagoon with *Rhizophora* and *Thalassia*; tidal flow; 0–1½ m.

Chl.: *Caulerpa sertularioides, Halimeda incrassata, H. opuntia, Penicillus capitatus.* — Rho.: *Melobesia farinosa, Spyridia filamentosa.*

1130A SIMSON BAY LAGOON, near former outlet, 6.VI.1955.

Lagoon shut off from sea recently; muddy pool with *Batophora*, dying *Rhizophora*; ¼–1 m (31.3 g Cl/l).

Chl.: *Batophora oerstedi.*

1131 SIMSON BAY LAGOON, LITTLE KEY, W. shore, 2.VIII.1949.

Muddy sand with some *Thalassia* and *Batophora*, on *Rhizophora*; 0–1½ m.

Chl.: *Acetabularia crenulata, Anadyomene stellata, Avrainvillea longicaulis, Batophora oerstedi, Caulerpa cupressoides, C. sertularioides, Halimeda incrassata, Valonia ventricosa.* — Phae.: *Dictyota cervicornis.* — Rho.: *Ceramium byssoideum, Champia parvula, Heterosiphonia gibbesii, Jania adhaerens, Laurencia intricata, L. obtusa, Melobesia farinosa, Polysiphonia ferulacea, Wrangelia argus.*

1132 SIMSON BAY LAGOON, W. shore of FLAMINGO POND, 8.VI.1949.

Muddy lagoon with rocky shore, nearly no open communication with sea; *Batophora* on *Rhizophora* and *Avicennia*; 0–1½ m.

Chl.: *Anadyomene stellata, Batophora oerstedi, Caulerpa sertularioides.* — Rho.: *Jania adhaerens, Polysiphonia ferulacea, Polysiphonia* spec., *Spyridia filamentosa.*

1132A SIMSON BAY LAGOON, FLAMINGO POND, 8.VI.1949.

Muddy lagoon with rocky shore, with small *Thalassia* and *Batophora*; ½–1½ m.

Chl.: *Anadyomene stellata, Batophora oerstedi.* — Rho.: *Melobesia farinosa.*

1132a SIMSON BAY LAGOON, FLAMINGO POND, 27.VI.1955.

Muddy lagoon, shut off from the sea, with *Rhizophora, Thalassia* and *Batophora*; 0–½ m (32.1 g Cl/l).

Chl.: *Acetabularia crenulata.* — Rho.: *Melobesia farinosa.* — Cya.: *Coccochloris elabens, Entophysalis deusta, Spirulina subsalsa.*

1133 ATWELL'S POND, S. corner, E. of Philipsburg, 17.V.1949.

Sheet of water on mud, 100 x 25 x ½ m, separated from sea by wall of debris, after heavy rains communicating with Roland's Canal and discharging into the sea; pieces of coral rock with *Enteromorpha*; clear water (31 g Cl/l).

Chl.: *Enteromorpha intestinalis.*

1399 POINT BLANCHE BAY, 5.VI.1955.

Surfswept limestone cliff with rock pools, 0–½ m.

Chl.: *Cladophoropsis membranacea.* — Phae.: *Dilophus alternans, Padina sanctae-crucis, Sargassum* spec., *Turbinaria turbinata.* — Rho.: *Digenea simplex, Laurencia microcladia, L. papillosa, Polysiphonia ferulacea, Spyridia aculeata.* — Cya.: *Dichothrix penicillata.*

1400   FRESHWATER POND, N.W. of Philipsburg, at bridge, 25.VII.1955.
       Shallow mudflat with *Ruppia* and algae; 0–½ m (25.5 g Cl/l).
       Chl.: *Cladophora crispula, Enteromorpha* spec.

1401   SIMSON BAY LAGOON, village, near jetty, 8.VI.1955.
       Muddy bottom among *Rhizophora* with *Syringodium* and *Batophora*; ½–1 m
       (29.1 g Cl/l).
       Chl.: *Batophora oerstedi, Enteromorpha prolifera.*

1402   SIMSON BAY LAGOON, near FLAMINGO POND, 27.VI.1955.
       Sandy shore with miserable *Rhizophora*, scanty *Thalassia* and *Syringodium*,
       0–½ m (28.4 g Cl/l).
       Chl.: *Batophora oerstedi.*

1403   SIMSON BAY LAGOON, FLAMINGO POND, entrance, 27.VI.1955.
       Almost isolated muddy pool among *Rhizophora* with *Batophora*; 0–½ m.
       Chl.: *Batophora oerstedi.*

ST. THOMAS

s.n.   CHARLOTTE AMALIE, harbour, 16.III.1937.
       On piece of rock in tidal zone.
           Chl.: *Ulva* spec. – Rho.: *Centroceras clavulatum, Gymnogongrus tenuis.*

114

CHAPTER VIII

BIBLIOGRAPHY

AGARDH, C. A., 1821–1828. *Species algarum rite cognitae, cum synonymis, differentiis specificis et descriptionibus succinctes.* Vol. *1, i* + 531 pp., 1821; vol. 2, lxvi + 189 pp., 1828. Greifswald.

AGARDH, J. G. 1847. Nya algar från Mexico. *Öfvers. Kongl. Vetensk.-Akad. Förhandl. 4,* p. 5–17.

AGARDH, J. G., 1848–1901. *Species, genera et ordines algarum, seu descriptiones succinctae specierum, generum et ordinum, quibus algarum regnum constituitur.* Vol. *1–3.* Lund.

ALMODÓVAR, L. R. & BIEBL, R., 1962. Osmotic resistance of mangrove algae around La Parguera, Puerto Rico. *Revue Algol.* (n.s.) 6, p. 203–208, 1 table, 1 map.

ARNOLDO, FR. M. (A. N. BROEDERS), 1954. *Wat in het wild groeit en bloeit op Curaçao, Aruba en Bonaire. Zakflora.* Uitgaven Natuurwet. Werkgroep Ned. Ant. *4;* Nijhoff, 's-Gravenhage, 170 pp., 68 pls. – 1964. *2nd. edition,* Uitg. Natuurwet. Werkgroep N. A. *16,* 229 pp., 68 pls.

BIEBL, R., 1962. Protoplasmatisch-ökologische Untersuchungen an Mangrovealgen von Puerto Rico. *Protoplasma 55,* p. 572–606, 17 figs.

BIEBL, R., 1962. Temperaturresistenz tropischer Meeresalgen. *Botanica marina 4,* p. 241–254, 1 fig., 3 tables.

BØRGESEN, F., 1900. A contribution to the knowledge of the marine alga vegetation on the coasts of the Danish West-Indian Islands. *Botanisk Tidsskr. 23,* p. 49–57.

BØRGESEN, F., 1909. Notes on the shore vegetation of the Danish West-Indian Islands. *Botanisk Tidsskr. 29,* p. 201–259, 40 figs., pl. III–IV.

BØRGESEN, F., 1911. The algal vegetation of the lagoons in the Danish West Indies. *Biologiske Arbejder til Eug. Warming,* p. 41–56, 9 figs. København.

BØRGESEN, F., 1913–1920. The marine algae of the Danish West Indies. I. Chlorophyceae. *Dansk Bot. Arkiv 1,* 1913, p. 1–160, fig. 1–126. II. Phaeophyceae. *ibid. 2,* 1914, p. 1–68, fig. 1–44. III. Rhodophyceae. *ibid. 3,* 1915, p. 1–80, fig. 1–86. *ibid. 3,* 1916, p. 82–144, fig. 87–148. *ibid. 3,* 1917, p. 145–240, fig. 149–230. *ibid. 3,* 1918, p. 241–304, fig. 231–307. *ibid. 3,* 1919, p. 305–368, fig. 308–360. *ibid. 3,* 1920, p. 369–504, fig. 361–435; 1 map.

BØRGESEN, F., 1930. Marine algae from the Canary Islands. Vol. 3, part 3. Ceramiales. *Kgl. Danske Vidensk. Selsk., Biol. Med. 9,* p. 1–159, fig. 1–60.

BRAAK, C., 1935. *Het klimaat van Nederlandsch West-Indië.* Meded. Verh. Nederl. Meteorol. Inst. *36,* 120 pp., 15 figs.

BRAUN-BLANQUET, J., 1964. *Pflanzensoziologie,* ed. 3, xiv + 865 pp., 442 figs., 88 tables. Wien.

BUISONJÉ, P. H. DE & ZONNEVELD, J. I. S., 1960. De kustvormen van Curaçao, Aruba en Bonaire. *Nieuwe West-Indische Gids 40,* p. 121–144, 16 phot., 7 figs.

CHAPMAN, V. J. & TREVARTHEN, C. B., 1953. General schemes of classification in relation to marine coastal zonation. *J. Ecol. 41,* p. 198–204, 2 tables.

CHRISTMAN, R. A., 1953. Geology of St. Bartholomew, St. Martin, and Anguilla, Lesser Antilles. *Bull. Geol. Soc. America. 64*, p. 65–96, 4 figs., 4 tables, 3 pls., 2 maps.

COKER, R. E. & GONZALEZ, J. G., 1960. Limnetic Copepod populations of Bahía Fosforescente and adjacent waters, Puerto Rico. *J. Elisha Mitchell Scient. Soc. 76*, p. 8–28, 4 figs., 5 tables.

COLLINS, F. S., 1901. The algae of Jamaica. *Proc. Amer. Acad. Arts Sci. 37*, p. 231–270, 6 tables.

COLLINS, F. S. & HERVEY, A. B., 1917. The algae of Bermuda. *Proc. Amer. Acad. Arts Sci. 53*, p. 1–195, pls. 1–6.

COMPÈRE, P., 1963. The correct name of the Afro-American black mangrove. *Taxon 12*, p. 150–152.

CONOVER, J. T., 1964. The ecology, seasonal periodicity, and distribution of benthic plants in some Texas lagoons. *Botanica marina 7*, p. 4–41, 6 figs., 3 tables.

COOMANS, H. E., 1958. A survey of the littoral Gastropoda of the Netherlands Antilles and other Caribbean Islands. *Studies fauna Curaçao 8*, Uitg. Natuurwet. Studiekring Sur. en Ned. Ant. *17*, p. 42–111, pl. I–XVI.

DAHL, E., 1953. Some aspects of the ecology and zonation of the fauna of sandy beaches. *Oikos 4*, p. 1–27, 8 figs.

DAVIS, J. H., 1940. The ecology and geologic rôle of mangroves in Florida. *Papers Tortugas Laboratory 32*, p. 303–412, 7 figs., 12 pls.

DE TONI, G. B., 1889–1924. *Sylloge algarum omnium hucusque cognitarum.* Vol. *1–6.* Padua.

DÍAZ-PIFERRER, M., 1964. Adiciones a la flora marina de las Antillas holandesas, Curazao y Bonaire. *Carib. J. Sci. 4*, p. 513–543, 25 figs.

EKMAN, S., 1953. *Zoogeography of the sea.* 417 pp., 121 figs., 49 tables. London.

ELLENBERG, H., 1956. *Aufgaben und Methoden der Vegetationskunde*, Teil 1, 136 pp., 19 figs., 21 tables, in: WALTER, *Einführung in die Phytologie*, Band IV. *Grundlagen der Vegetationsgliederung.* Stuttgart.

FELDMANN, J., 1946. La flora marine des Iles Atlantides. *Mém. Soc. Biogeogr. 8*, p. 395–435.

FELDMANN, J., 1951. Ecology of marine algae, p. 313–334, in: SMITH, *Manual of Phycology*, 375 pp., Waltham (Mass.).

FELDMANN, J., 1955. La zonation des algues sur la côte atlantique du Maroc. *Bull. Soc. Sc. Nat. Phys. Maroc 35*, p. 9–17, 1 table.

FELDMANN, J. & LAMI, R., 1936. Sur la végétation de la mangrove à la Guadeloupe. *Compt. Rend. Acad. Sci.* Paris, *203*, p. 883–885.

FELDMANN, J. & LAMI, R., 1937. Sur la végétation marine de la Guadeloupe. *Compt. Rend. Acad. Sci.* Paris, *204*, p. 186–188.

FRÉMY, P., 1941. Cyanophycées des îles Bonaire, Curaçao et Aruba d'après les récoltes de M. Wagenaar Hummelinck (Utrecht) en 1930. *Revue Algol. 12*, p. 101–152. See also: *West-Indische Gids 27*, 1944, p. 62–64.

GAULD, D. T. & BUCHANAN, J. B., 1956. The fauna of sandy beaches in the Gold Coast. *Oikos 7*, p. 293–301.

GERLACH, S. A., 1958. Die Mangroveregion tropischer Küsten als Lebensraum. *Z. Morphol. Ökol. Tiere 46*, p. 636–730, 25 figs., 20 tables.

GUILER, E. R., 1953. Intertidal classification in Tasmania. *J. Ecol. 41*, p. 381–384.

116

GUILER, E. R., 1955. Australian intertidal belt-forming species in Tasmania. *J. Ecol.* *43*, p. 138–148, 1 table.

HAAN, D. DE & ZANEVELD, J. S., 1959. Some notes on tides in Annabaai Harbour, Curaçao, Netherlands Antilles. *Bull. Mar. Sci. Gulf Caribb. 9*, p. 224–236, 7 figs., 5 tables.

HAMEL, G. & HAMEL-JOUKOV, A., 1929–1931. *Algues des Antilles Françaises* (exsiccata). Fasc. 1–3, Paris.

HARTOG, C. DEN, 1959. The epilithic algal communities occurring along the coast of the Netherlands. *Wentia 1*, p. 1–241. 27 figs., 43 tables.

HARVEY, W. H., 1852–1858. Nereis Boreali-Americana. I, Melanospermae. *Smiths. Contrib. to Knowledge 3*(4), p. 1–150, pl. 1–12, 1852; II, Rhodospermae, *ibid.* 5(5), p. 1–258, pl. 13–36, 1853; III, Chlorospermae, including supplements, *ibid.* *10*, p. ii + 1–140, pl. 37–50, 1858.

HOWE, M. A., 1920. Algae, in: BRITTON & MILLSPAUGH, *The Bahama Flora*, p. 553–618. New York.

HUMM, H. J., 1963. Algae of the Southern Gulf of Mexico. *Proc. 4th Intern. Seaweed Symposium*, p. 202–206. London.

HUMM, H. J., 1964. Epiphytes of the sea grass, Thalassia testudinum, in Florida. *Bull. Mar. Sci. Gulf Caribb. 14*, p. 306–341, 3 figs.

JAASUND, E., 1965. Aspects of the marine algal vegetation of North Norway. *Acta Universitatis Gothoburgensis*, p. 1–174, 43 figs.

KJELLMAN, F. R., 1878. Ueber Algenregionen und Algenformationen im östlichen Skagerrak. *Bih. Kgl. Svenska Vet. Akad. Handl. 5*, p. 1–35.

KORNAŚ, J. & MEDWECKA-KORNAŚ, A., 1959. Associations végétales sous-marines dans le Golfe de Gdańsk (Baltique polonaise). *Vegetatio 2*, p. 120–127, 3 tables, 1 map.

KORNAŚ, J. & PANCER, E. & BRZYSKI, B., 1960. Studies on sea-bottom vegetation in the Bay of Gdańsk off Rewa. *Fragm. Flor. Geobot. 6*, p. 1–92, 31 figs., 12 tables.

KOSTER, J. T., 1943. Some Chlorophyceae from the marine salines of Bonaire (Netherlands West Indies). *Blumea 5*, p. 328–335, 2 figs.

KOSTER, J. T., 1960. Caribbean brackish and freshwater Cyanophyceae. *Blumea 10*, p. 323–366, 78 figs., 1 map.

KOSTER, J. T., 1963. Antillean Cyanophyceae from salt-pans and marine localities. *Blumea 12*, p. 45–56.

KYLIN, H., 1956. *Die Gattungen der Rhodophyceen.* xv + 673 pp., 458 figs. Lund.

LAWSON, G. W., 1956. Rocky shore zonation in the Gold coast. *J. Ecol. 44*, p. 153–170, 6 figs., 1 pl.

LAWSON, G. W., 1957. Seasonal variation of intertidal zonation on the coast of Ghana in relation to tidal factors. *J. Ecol. 45*, p. 831–860, 9 figs., 19 tables.

LEWIS, J. B., 1960. The fauna of rocky shores of Barbados, West Indies. *Canad. J. Zool. 38*, p. 391–435, 20 figs.

LEWIS, J. R., 1955. The mode of occurrence of the universal intertidal zones in Great-Britain. *J. Ecol. 43*, p. 270–290, 4 figs.

MAZÉ, H. & SCHRAMM, A., 1870–1877. *Essai de classification des algues de la Guadeloupe.* xix + 283 + 3 pp. Basse Terre.

MOLINIER, R., 1960. Études des biocénoses marines du Cap Corse (France). *Vegetatio 9*, p. 121–192, 217–312; 21 figs. 6 phot.

117

MONTAGNE, J. F. C., 1842. Algae, p. 1—104, pl. 1—5, in: DE LA SAGRA, *Histoire physique, politique et naturelle de l'Ile de Cuba*, Botanique — plantes cellulares, x + 549 pp. (1838—1842). Paris.

MURRAY, G., 1889. Catalogue of the marine algae of the West Indian region. *J. Bot.* 27, p. 298—305, 5 tables.

NEWELL, N. D. & IMBRIE, J. & PURDY, E. G. & THURBER, D. L., 1959. Organism communities and bottom facies, Great Bahama Bank. *Bull. Amer. Mus. Nat. Hist.* 117, 4, p. 179—228, pl. 58—69, fig. 1—17, tables 1—6.

PARR, A. E., 1937. A contribution to the hydrography of the Caribbean and Cayman seas. *Bull. Bingham Oceanogr. Coll. 5*, 4, p. 1—110, 82 figs.

PARR, A. E., 1938. Further observations on the hydrography of the eastern Caribbean and adjacent Atlantic waters. *Bull. Bingham Oceanogr. Coll. 6*, 4, p. 1—29, 23 figs.

PATULLO, J. & MUNK, W. & REVILLE, R. & STRONG, E., 1955. The seasonal oscillation in sea level. *J. Mar. Res. 14*, p. 88—156, 5 figs., 3 tables, 2 maps.

POST, E., 1936. Systematische und pflanzengeographische Notizen zur Bostrychia-Caloglossa Assoziation. Ergebn. Sunda-Exped. 1929/30. *Revue Algol. 9*, p. 1—84, 4 figs.

QUESTEL, A., 1951. Algae, p. 193—208, in: *La flora de la Guadeloupe*. Basseterre.

RAKESTRAW, N. W. & SMITH, H. P., 1937. A contribution to the chemistry of the Caribbean and Cayman seas. *Bull. Bingham Oceanogr. Coll. 6*, 1, p. 1—41, 32 figs.

RODRIGUEZ, G., 1959. The marine communities of Margarita Island, Venezuela. *Bull. Mar. Sci. Gulf Carrib. 9*, p. 237—280, 26 figs.

ROSS, R. & IRVINE, L. M., 1967. The typification of the genus Byssus L. (1753). *Taxon 16*, p. 184—186.

SKOTTSBERG, C., 1941. Communities of marine algae in subantarctic and antarctic waters. *Kungl. Svenska Vetensk. Akad. Handl.* (3) *19*, nr. 4, p. 1—92, 7 figs., 3 pls.

SLUITER, C. P., 1908. List of algae collected by the Fishery Inspection at Curaçao. *Rec. Trav. Bot. Néerl. 4*, p. 231—241, pl. 8.

SMITH, C. L., 1940. The Great Bahama Bank. I. General hydrographical and chemical features. *J. Mar. Res. 3*, p. 147—170, fig. 44—51, 4 tables.

SOUTHWARD, A. J., 1958. The zonation of plants and animals on rocky sea shores. *Biol. Rev. 33*, p. 137—177.

Statistiek van de meteorologische waarnemingen van de Nederlandse Antillen 1957. Dienst Econom. Zaken Welvaartszorg, Bureau Statistiek Ned. Ant. 5, 27 pp. — 1958, 6, 34 pp.

STEPHENSON, T. A. & STEPHENSON, A., 1949. The universal features of zonation between tide marks on rocky coasts. *J. Ecol. 37*, p. 289—305, 4 figs., pl. 8.

STEPHENSON, T. A. & STEPHENSON, A., 1950. Life between tide marks in North America. I. The Florida Keys. *J. Ecol. 38*, p. 354—402, 10 figs., pls. 9—15.

STOFFERS, A. L., 1956. *The vegetation of the Netherlands Antilles.* Studies on the flora of Curaçao and other Caribbean Islands *1*. Uitg. Natuurwet. Studiekring Sur. en Ned. Ant. *15*, 142 pp., 12 figs., 28 pls., 4 col. maps.

TAYLOR, W. R., 1929. Notes on algae from the tropical Atlantic Ocean. I. *Amer. J. Bot. 16*, p. 621—630, 13 figs., pl. 62.

TAYLOR, W. R., 1940. Marine algae of the Smithsonian-Hartford Expedition to the West Indies, 1937. *Contr. U.S. Nat. Herb. 28*, p. 549—562, pl. 20.

TAYLOR, W. R., 1942. Caribbean marine algae of the Allan Hancock Expedition, 1939. *Rep. Allan Hancock Atlantic Exped.* 2, 193 pp., 20 pls.

TAYLOR, W. R., 1950. Marine algal flora of the Caribbean and its extension into neighbouring seas. *VIIIe Congrès Intern. Bot., Rapports,* p. 149–150.

TAYLOR, W. R., 1951. Survey of the marine algae of Bermuda. *Year Book Amer. Philos. Soc. 1951,* p. 167–171.

TAYLOR, W. R., 1954. Sketch of the character of the marine algal vegetation of the shores of the Gulf of Mexico. *Fishery Bull. 89,* p. 177–192, fig. 48–50.

TAYLOR, W. R., 1955. Marine algal flora of the Caribbean and its extension into neighbouring seas. *Essays in the natural sciences in honor of Captain Allan Hancock,* p. 259–270, 8 figs., 2 tables.

TAYLOR, W. R., 1959a. Associations algales des mangroves d'Amérique. *Colloques intern. centre national recherche scient. 81, Écologie des algues marines,* p. 143–152.

TAYLOR, W. R., 1959b. Phycology in retrospect and anticipation. *Vistas in Botany,* p. 328–347. New York.

TAYLOR, W. R., 1960. *Marine algae of the eastern tropical and subtropical coasts of the Americas.* ix + 870 pp., 14 figs., 80 pls. Ann Arbor.

TAYLOR, W. R., 1961. Distribution in depth of marine algae in the Caribbean and adjacent seas. *Recent advances in Botany, Ecology of marine algae,* p. 193–197, 1 table. Toronto.

TAYLOR, W. R., 1962. Marine algae from the tropical Atlantic Ocean. V. Algae from the Lesser Antilles. *Contr. U.S. Nat. Herb. 36,* p. 43–62, 4 pls.

VICKERS, A., 1905. Liste des algues marines de la Barbade. *Ann. Sci. Nat., Bot. 9,* p. 45–66.

VICKERS, A., 1908. *Phycologia barbadensis. Iconographie des algues marines récoltées à l'île Barbade (Antilles),* Part 1: Chlorophyceae, p. 1–30, 53 pls.; part 2: Phaeophyceae, p. 31–44, 34 pls. Paris.

VOSS, G. L. & VOSS, N. A., 1955. An ecological survey of Soldier Key, Biscayne Bay, Florida. *Bull. Mar. Sci. Gulf Caribb. 5,* p. 203–229, 4 figs.

VROMAN, M., 1959. Onderzoek naar het voorkomen en de oecologie van de mariene wieren der Nederlandse Antillen. *Jaarbericht WOSUNA 1958,* p. 38–39, 2 pls.

VROMAN, M., 1967. A new species of Stichothamnion (Rhodophyta) from the West Indies. *Acta Bot. Neerl. 15,* p. 557–561, 3 figs., 4 pls.

WAGENAAR HUMMELINCK, P., 1953. Description of new localities. *Studies fauna Curaçao 4;* Uitg. Natuurwet. Studiekring Sur. en Ned. Ant. 8, p. 1–108, 25 figs., 8 pls.

WESTERMANN, J. H., 1949. *Overzicht van de geologische en mijnbouwkundige kennis der Nederlandse Antillen.* Meded. Indisch Instituut Amsterdam 85, 168 pp., 10 figs., 24 phot.

WESTERMANN, J. H., 1957. De geologische geschiedenis der drie Bovenwindse Eilanden St. Martin, Saba en St. Eustatius. *West-Indische Gids 37,* p. 127–168, 11 figs., 16 phot.

WESTERMANN, J. H. & KIEL, H., 1961. *The geology of Saba and St. Eustatius, with notes on the geology of St. Kitts, Nevis and Montserrat.* Uitg. Natuurwet. Studiekring Sur. en Ned. Ant. 24; xiv + 175 pp., 11 figs., 33 pls., 6 col. maps.

WESTHOFF, V., 1951. An analysis of some concepts and terms in vegetation study or phytocoenology. *Synthese 8,* p. 194–206.

Womersley, H. B. S. & Edmonds, S. J., 1952. Marine coastal zonation in Southern Australia in relation to a general scheme of classification. *J. Ecol. 40*, p. 84—90, 1 fig.

Womersley, H. B. S. & Edmonds, S. J., 1958. A general account of the intertidal ecology of South Australian coasts. *Austr. J. Mar. Freshw. Res. 9*, p. 217—260, 2 figs., 2 tables, 12 pls.

Zaneveld, J. S., 1956. Het Caraïbisch Marien-Biologisch Instituut te Curaçao. *Vakblad Biologen 36*, p. 249—259, 4 figs.

Zaneveld, J. S., 1958. A Lithothamnion bank at Bonaire. *Blumea, suppl. 4*, p. 206—219, 6 figs.

CURRICULUM VITAE

De auteur werd geboren te Woerden op 28 februari 1927. In deze plaats ontving hij ook zijn vooropleiding. Na het verkrijgen van het diploma HBS-B werd in januari 1946 begonnen met de studie in de biologie aan de Rijksuniversiteit te Utrecht. Het candidaatsexamen werd afgelegd op 6 juli 1948.

Na het vervullen van de militaire dienstplicht van september 1948 tot december 1949 werd de studie in 1950 hervat.

Na tijdelijke leraarsfuncties te hebben vervuld in Rotterdam en Arnhem werd hij per 1 september 1951 benoemd tot assistent aan het Botanisch Laboratorium van de toen juist opgerichte afdeling Biologie aan de Vrije Universiteit te Amsterdam en speciaal belast met de leiding van de practica voor het vak plantensystematiek. Het doctoraalexamen werd afgelegd te Utrecht op 5 juli 1955. Op 1 october 1955 volgde een benoeming tot wetenschappelijk ambtenaar aan de Vrije Universiteit te Amsterdam, met leeropdracht voor de plantensystematiek.

Met steun van de Stichting voor Wetenschappelijk Onderzoek in Suriname en de Nederlandse Antillen (WOSUNA) werd vanaf november 1957 tot juli 1958 in de Nederlandse Antillen onderzoek verricht naar de aldaar voorkomende mariene wiersoorten. Een deel van de resultaten van de bewerking van het toen verzamelde materiaal wordt in deze publicatie samengevat.

PLATE I

Ia.   St. Martin: the peninsula of Point Blanche, as seen towards the south, with
Point Blanche Bay.

Ib.   St. Martin: Lay Bay, with rocks of the Low Lands formation.

PLATE II

IIa. St. Martin: looking from Mary Point eastward, with cliffs of the Low Lands formation.

IIb. St. Martin: view of Mary Point, with cliffs of the Low Lands formation.

PLATE III

IIIa.　St. Martin: dolerite rocks at the eastern side of the peninsula on which Fort Amsterdam has been built.

IIIb.　St. Martin: view of Little Bay (left) and Great Bay (right) from Fort Amsterdam.

PLATE IV

IVa.   St. Martin: Guana Bay, with Pleistocene coral limestone in the southern part.

IVb.   St. Martin: zonation of organisms on coral limestone at Point Blanche Bay.

PLATE V

Va.   ST. MARTIN: Cole Bay as seen from its western corner.

Vb.   ST. MARTIN: eastern shore of Simson Bay, with rocks of the
Low Lands information.

PLATE VI

VIa. St. Martin: sandy beach with beachrock in the western part of Burgeux Bay.

VIb. St. Martin: western part of Maho Bay, with rocks of the Low Lands formation, showing a niche.

PLATE VII

VIIa. St. Martin: Little Bay near Kay Bay Hill, with rocks of the
Point Blanche formation.

VIIb. St. Martin: view of Great Bay towards Fort Hill, as seen from the landing-
stage, with boulders of the Point Blanche formation.

PLATE VIII

VIIIa. St. Martin: sandy beach with rocks of the Point Blanche formation at Oyster Pond.

VIIIb. St. Martin: Guana Bay; rocks of the Point Blanche formation, partly covered by gently dipping coral limestone.

PLATE IX

IXa. ST. MARTIN: Baie de la Grande Case; a sandy beach with beachrock, with a coast of dioritic boulders at the background.

IXb. SABA: andesitic rocks in the tidal zone at Fort Bay.

PLATE X

X. St. Eustatius: limestone cliff of Sugarloaf and White Wall, with a boulder
beach of andesitic rock.